D1417827

TOWARD POLLUTION-FREE MANUFACTURING

Institute for
Local Self-Reliance

AMA Management Briefing

AMA MEMBERSHIP PUBLICATIONS DIVISION
AMERICAN MANAGEMENT ASSOCIATION

Library of Congress Cataloging-in-Publication Data

Toward pollution-free manufacturing.

(AMA management briefing)
Bibliography: p.
1. Factory and trade waste—Environmental aspects—
United States. 2. United States—Manufactures.
I. Institute for Local Self-Reliance. II.Series.
TD897.7.T69 1987 363.7'23 86-17424
ISBN 0-8144-2327-2

2425 18th Street, N.W., Washington, D.C. 20009 (202) 232-4108

This Management Briefing has been distributed to all members enrolled in the Manufacturing and Research and Development Divisions of the American Management Association. Copies may be purchased at the following single copy rates: AMA members, $7.50. Nonmembers, $10.00. Students, $3.75 (upon presentation of a college/university identification card at an AMA bookstore) Faculty members may purchase 25 or more copies for classroom use at the student discount rate (order on college letterhead).

First Printing

Acknowledgments

The Institute for Local Self-Reliance (ILSR) is a nonprofit research, public policy, and technical assistance organization which, since 1974, has helped dozens of communities achieve greater self-sufficiency by incorporating sound waste utilization minimization and job development strategies into their economic development programs. Currently, ILSR works with state and local governments, industrial firms, small businesses, and civic and community organizations.

Much of the information included in this publication was adapted from the book, *Proven Profits from Pollution Prevention,* authored by Don Huisingh, Larry Martin, Helene Hilger, and Neil Seldman, and published by the Institute for Local Self-Reliance. The 316-page volume can be ordered for $26.50 (prepaid) from the Institute for Local Self-Reliance, Department A, 2425 18th Street, N.W., Washington, D.C. 20009, (202) 232-4108.

The editorial content of this publication represents a cooperative effort between the Institute for Local Self-Reliance and the AMA editorial staff. AMA staff writer John Aberth served as project director, editor, and reporter.

Contents

Introduction

t was lunch time at the Office of Technology Assessment's workshop
n hazardous waste avoidance. More than a dozen businessmen from
mall and medium-sized companies, most of them CEOs, were sitting
round eating lasagna. I happened to find myself sharing a table with a
roup of electroplaters. One of them, reminded by the lasagna of his
azardous waste, started explaining how he got rid of his noxious brew:

"I have a treater come and haul off my hazardous sludge. He takes it
ack to his plant and mixes it with lime, or something. He says this
eutralizes it so he can send it to the local landfill as ordinary waste.
hat's the advantage of this treater over the others: He'll charge me
ower 'cause he doesn't have to use the hazardous landfill.

"Once I give him the drums, I'm done with it. My name is off and I
on't have to worry about it anymore."

He sat back, a contented smile across his face. Joel Hirschhorn,
roject director for the OTA and leader of the workshop, spoke up.
You're completely wrong, you're not done with it," he said heatedly.
In fact, you are liable for *two* sites: the plant where he treats the waste
nd the landfill that he sends it to."

The plater's face dropped. His voice had a new tone of concern.
Really?" he said in disbelief. "I thought that once he took it away, that
vas it, the waste was off my hands."

Joel Hirschhorn was vigorously shaking his head. "I'm afraid it
oesn't work that way. The generator is liable no matter how the waste
s treated or where it goes. And you're liable for two sites. I'd be worried
I were you."

"Believe me, I am," was the plater's reply.

Labelled containers of treated, hazardous waste awaiting removal to a disposal facility.

THE PIONEERS SAY IT LOOKS GOOD

The conversation is typical of the many candid exchanges that toc place during the course of the two-day workshop. Industry-specif meetings such as these come as one prong of the OTA drive to prepa for congressional hearings set for October of this year. The Enviro mental Protection Agency is engaged in a parallel fact-finding effort, ar the reports and legislative recommendations of these two agencies ma herald a new era in industry's approach to hazardous waste manag ment.

The technologies now exist for industry to dramatically reduce and in some cases, eliminate—hazardous waste discharged into th environment. The long-term implications are enormous.

Given enough time and the appropriate assistance, the waste r duction concept could spell an end to Superfund, build an escape hatc for manufacturers trapped in the liability insurance spiral, calm th rancor of environmentalists, and produce a cleaner environment fo millions of Americans.

Like any new idea, the waste reduction movement will have its evangelists as well as its cynics and nay-sayers. The pivitol decision point, of course, is with the manufacturer. Interestingly, the forecast from those who have already implemented waste-reduction programs is, indeed, bullish. The pioneers are reporting that yes, there are pitfalls, but the idea makes sense, for a variety of reasons, economic as well as environmental.

The metalplaters' workshop is a case in point. Let's take a closer look at that meeting, as a point of departure for the following chapters.

"BURNED FOR $3.5 MILLION"

The participants, representatives of small-to-medium sized businesses (the large-company representatives had met earlier in the week), gathered in one of the OTA's conference rooms at 9 o'clock in the morning. The day started off with everyone telling how they got into hazardous waste management. Various reasons, some of them quite revealing, emerged.

"I've been involved with the plating industry for over 10 years," said David Anzures of California Technical Plating, "and I saw what was coming. I learned enough about the environment, and about the legislation, not to fall into that problem. I didn't want to be a sludge producer."

Another representative, who wishes to remain anonymous, mentioned a very big reason for his concern with waste avoidance: "Our company got burned for 3 1/2 million dollars to clean up a Superfund site. The man responsible for disposing our wastes dumped them into a

---

People were noticing that their water was tasting funny. In fact, they were afraid to light a match near it.

---

lagoon instead of treating them properly, as he should. People were noticing that their water was tasting funny. In fact, they were afraid to light a match near it.

"This joker who burned us later went on to convince a farmer to accept waste in his backyard in exchange for so many gallons of paint.

That farmer now has a freshly painted Superfund site."

"At one point, 50 percent of our production people had dermatitis said Paul Hoffman of Garden Way, Inc. His concern for his employee health led him to investigate ways to reduce the toxicity of his produ tion line.

"One of our considerations in looking at waste reduction is that it's marketing decision," said Richard Herring of The Gloucester Cor pany. "People who are not used to adhesives will generally shun labe that say, 'Might be fatal.'"

"We'd all like to say we're for motherhood and America," said Har DeSoi of Pioneer Metal Finishing. "Hey, I live for the Lord. But it's th regulations that often force me into waste reduction."

"It's all based on economics for the small businessman . . . Well-managed businesses are putting in waste reduction because they want to stay in business."

"It's all based on economics for the small businessman," said Co Platt of Precious Metal Platers. "I looked at cadmium plating, fo example. I saw it was going to be hazardous and I didn't go into i Well-managed businesses are putting in waste reduction because the want to *stay* in business."

REGULATION: PART OF THE PROBLEM, OR PART OF TH SOLUTION?

The discussion turned toward how successful each company had bee in implementing its waste reduction strategy. Various regulatory age cies, particularly the EPA, become the target of criticism.

"Often it's not the desire to be unregulated that motivates us in th battle," said David Anzures. "I want a clean plant for myself and m people. Most electroplaters would've welcomed regulation if they had system for recovery and reclamation. But this system of regulatio which brings on more regulation, that says we have to do it this wa instead of that way, isn't helping us any."

"The EPA was threatening to close me down, just as I was begin-
ing to clean up my hazardous waste," said Harry DeSoi. "They helped
ne take out a loan to finance my project, but then they put me on a
trict timetable to come into compliance. When my time was up, forget
, they wouldn't have given me a second chance."

"We're out in the sticks; we don't have the help that most companies
ave," said Robert Hall of ESAB Health, Inc. "It's very time consuming
o research and develop waste reduction. And the EPA is causing me
nore problems than it's solving. The laws are changing so fast that it's
ard to keep up with all of them."

Many agreed, claiming that even their nonhazardous sludge is being
ent to a hazardous landfill, since they don't know what's going to be
lassified as hazardous two years down the road.

Regulations are extremely complicated and chancy. All you have to do is hiccup and you're out of compliance.

"Regulations are extremely complicated and chancy," said Rudolf
Gabel of Syborn Corporation. "All you have to do is hiccup and you're
out of compliance. Everyone in industry knows that wastes are bad.
You have to pay for them. But regulations don't take into consideration
ll the efforts that industry has made."

Phil Horelick of Allied Metal Finishing has put much time and effort
nto trying to comply with regulations. "Look at this, it's a complete
ightmare," he said as he handed out photos of his waste-water treat-
nent system (see next page). "It all looks very exotic and showboaty,
ut it's very difficult to maintain. I had to put it in to stay in business, to
e in compliance. But the machine stays in compliance only if it's very
losely attended to. Plus it's very expensive to buy and operate.

I set up a display where I have fish living in the treated water next to a bottle full of the original sludge.

Control panel regulating the chemical reactions in Allied's waste treatment facility.

Containers in which the chemical reagents neutralize specifically targeted waste streams.

"The water that comes out of it though is actually very impressive. I et up a display where I have fish living in the treated water next to a ottle full of the original sludge.

"Everything I've done is waste reduction. It's absolutely essential. ve proved it; I'm one of the survivors, one of the few left in my area. But m finding that more and more of my time and my plant space are being aken up with all this equipment. It's the Dr. Jekyll and Mr. Hyde yndrome—I'm changing from a plater into a full-time, hazardous waste eater!"

Horelick's frustration is echoed by many in the industry, who find at the strict timetables of regulations often force them to put in waste eatment systems which, in fact, do not accomplish any reduction of aste. In Horelick's case, the treatment system merely removes waste om the water and concentrates it into a sludge for land disposal. echnically, this is *not* waste reduction.

Paul Hoffman was confused by what the Resource Conservation nd Recovery Act (RCRA) means when it asks generators to certify at they have a formal waste reduction program in place. "What *is* a rmal program? Is everyone wearing tuxedos?!" he asked.

Another CEO put forward his own reasons for distrusting RCRA nd the EPA: "I think RCRA is just an excuse used by the EPA to create ore paperwork so that they can keep their jobs. After all, if the EPA *eally* got us to eliminate all of our waste, wouldn't that put them out of usiness?"

Others were concerned about the legal liability that seems to per-anently attach itself to anyone producing hazardous waste. "Even if ou've sold your company and you're retired in Florida, they *still* can ome after you!" quipped one CEO.

There were complaints against other parties besides the EPA, of ourse. One of the targets was large companies: "A lot of the big guys ride themselves on saying that they accept products only from EPA-

I can plate a bolt pollution-free, but it's going to cost $2.35 instead of 35¢. The government buys 5 million cadmium-plated bolts per day. Which one do you think they're going to buy?

approved companies," said Harry DeSoi. "But it's just not true. IBM uses companies that are not in compliance, just because they're cheaper. Those who *are* in compliance, like me, can't compete."

David Anzures had a similar complaint about the government: "I can plate a bolt pollution-free, but it's going to cost $2.35 instead of 35¢. The government buys 5 million cadmium-plated bolts per day. Which one do you think they're going to buy?

"I am also bound by their specs. If I have a contract with the military, for example, I cannot use a different process in order to avoid waste. I have to use a cadmium-cyanide-based bath, not an acid bath. I have no choice in the matter. That's government regulation."

IF REGULATION, THEN OBLIGATION

But after the complaints, the industry representatives had some positive suggestions as to how the government could help them implement waste reduction programs. Said David Anzures, "Once government has stepped in and regulated, it has nothing else but a moral obligation to help those that are being regulated.

"I need some relaxation of regulation—a variance procedure for those engaged in recycling or recovery. I don't need money; I need cooperation by government agencies, and their legislation, in this process. Waste avoidance, if it's done properly, really doesn't cost anybody any money."

The participants mentioned other sources that proved helpful to them: State universities were useful in communicating the technology of waste reduction. The Small Business Administration (SBA) enabled many companies to take out loans in order to finance their waste reduction projects.

It's very disappointing to have the EPA spend millions of dollars to tell us something we already know.

But perhaps the most valuable help of all will come from industry itself. Both George McRae of Stanadyne, Inc. and David Anzures found that trade associations, such as The American Electroplaters' Society, were especially helpful in providing valuable information. David Anzures argues, "The only kind of education that we electroplaters pay attention to is from our peers. It's very disappointing to have the EPA spend millions of dollars to tell us something we already know." Others added that many consultants and regulatory agencies don't have the technical expertise to be really helpful.

"It's amazing how freely information is exchanged," said Dennis Drolet of Norchem. "It's a saying in industry that we have to help each other. Of course there are limits where a company's trade secrets are concerned. But there's an awful lot of knowledge you can exchange without giving away proprietary information."

We should all hang together or we'll be hung separately.

Cort Platt, who formed an association of printed circuit manufacturers and plating shops in order to pool their resources in waste reduction, said it best: "We should all hang together or we'll be hung separately."

Finally, we should let Jim Hornberg of The Dexter Corporation have the last word of advice about the alternative to implementing waste reduction: "If you close down your operation, you don't make anything and you don't produce any waste. That's waste avoidance!"

ABOUT THE FOLLOWING CHAPTERS

Those who are skeptical about waste reduction will continue to cite such matters as unsupportive regulations, unreliable machinery, and high costs (especially for smaller manufacturers) as roadblocks. And those who champion waste reduction will have to address these obstacles head on.

In discussing the role of the EPA and other agencies, the electroplaters suggested a number of ways in which government could play a supportive role. These included:

- Blacklisting of suppliers not in compliance with RCRA.
- A federal tax credit for waste reducers.
- A longer permit life for treaters.
- That insurance companies take notice of EPA-certified "goo actors" in waste management and reduce their premiums accorc ingly.

Chapter 1 puts these suggestions, as well as other questions, to EP director Lee Thomas.

After surveying the procedures and technologies involved in wast reduction (Chapter 2), we turn to a series of case studies showing ho various firms have fared in their waste reduction programs. The cas studies (Chapters 3 to 7) focus primarily on companies that hav achieved significant returns on their waste reduction investments. W offer these studies as examples of successful strategies or technologie that can be adapted to other locations. Many are low cost, and all hav been documented to show a reasonable return on investment.

Finally, Chapter 8 addresses the financial implications of starting waste reduction program.

We hope this briefing will provide some answers to industry's search fc ways to reduce, even eliminate, its hazardous waste production. Oth sources with more detailed technical information can be found in th bibliography and contact lists at the end of the chapters 3, 4, 5, 6, and

JOHN S. ABERTH
Editor

1

Meet Mr. Lee Thomas

Next to a shopping center, somewhere in southwest Washington D.C., rises a monolithic mass of concrete and glass, looking something like the apartment buildings across the street. This is the West Tower of the Environmental Protection Agency.

Up on the top floor, a spacious office offers a panoramic view of the city. On either side of the door are shelves reaching up to the ceiling.

Two things immediately stand out on the shelves. One is a bottle apparently full of wine. A close inspection of the label, however, reveals its true contents: "Vintage Acid Rain, 1979."

The other object is a construction hat with a name across the front in bold, black letters: **Lee Thomas.**

Then there is Thomas himself, the latest in the line of EPA Administrators, his face and figure looking surprisingly relaxed and contented. His Southern drawl is pleasant to hear. "Ya'll doin' a publication on waste reduction?" he asks.

"Yes," you reply. "Industry leaders have expressed concern about the Resource Conservation and Recovery Act—its strict regulations and lack of incentives. What are your thoughts on the matter?"

Thomas, who must report to Congress in October when the lawmakers reauthorize RCRA, fields the question effortlessly. You listen. The interview proceeds.

Lee Thomas: My impression is that waste reduction is important and going to become more important for industry to take a look at as far as their overall practices for managing hazardous wastes. Our report is going to reinforce that view, at least from what I've seen so far. It's going to be structured so that there are a lot of case examples of where waste reduction has had an impact on industry, and how it has been put in place.

There's going to be a lot of analyses of how the concept of waste reduction could be incorporated into a regulatory scheme, and whether it *should* be incorporated into a regulatory scheme. The evidence is mounting that waste reduction can have a major impact both on how much money an industry spends on hazardous waste management, and how much protection the environment gets.

John Aberth: Do you think waste reduction will make economic development and environmental quality compatible?

LT: Yes, I see that happening now. I believe that environmental protection and economic development can be made compatible, and they are compatible when you look at the case studies that have been done on waste reduction.

Look at some of those studies. You'll see companies saying, 'If only we'd looked at this a few years ago, look how much money we would've saved.'

Well, from our point of view, if only they had looked at it a few years ago, look what kind of protection would've accrued. We wouldn't have this waste in a landfill that we're now cleaning up.

Waste reduction is a perfect example to me of where the economics and the environment can be very compatible.

JA: Do you see a new era in hazardous waste management where industry becomes the new champion of a

Lee M. Thomas was designated by President Reagan to succeed William D. Ruckelshaus as Administrator of the Environmental Protection Agency on November 29, 1984, and he was named Acting Administrator on January 6, 1985. He was confirmed by the Senate on February 6 and sworn in as Administrator on February 8, 1985.

Previously, Mr. Thomas served as Assistant Administrator for Solid Waste and Emergency Response, advising the Administrator on methods of managing solid waste and hazardous material in such a way as to safeguard the public health and property. He was responsible for implementing the Resource Conservation and Recovery Act (RCRA), and for controlling presently operating waste sites and the Comprehensive Environmental Response Compensation and Liability Act (CERCLA), which established a $1.6 billion Superfund to clean up abandoned hazardous waste dumps.

Mr. Thomas was born in Ridgeway, South Carolina on June 13, 1944. He received his Bachelor of Arts degree in psychology from the University of the South, Sewanee, Tennessee and a Master of Arts in Education from the University of South Carolina, where he also studied postgraduate psychology. He currently resides in Woodbridge, Virginia with his wife Dixie and their two sons, Jordan and Braden.

---

clean environment through aggressive waste reduction practices?

LT: Yes, to a certain extent I do see that. As a matter of fact, about a year ago I saw an international environmental group give out an award to 3M [the Minnesota Mining and Manufacturing Company] for their "Pollution Prevention Pays" program. That's as good an example as I've seen in a long time of one industry standing before other industries and saying, 'Look how much sense this made to us in terms of the money

we've saved, as well as complying with environmenta
regulations.'

Look at government's role in the management c
hazardous waste. There are a number of things hap
pening that are going to get industry more and more i
tune with the waste reduction or recycling effort. O
the other hand, the regulatory side under RCRA i
getting tighter and tighter as to what a company ca
do with its waste, where it can dispose of that waste
and under what conditions. The cost of disposal i
increasing; just from an economic view, it's going t
drive industry to look harder at the generation c
waste in the first place, and it's going to raise th
bottom line to spend more money in the waste reduc
tion-recycling process.

On the other side of the coin are the liability con
cerns that industry is beginning to realize. As we'v
become increasingly aggressive in implementing th
Superfund program, industry is becoming more an
more aware of the long-term liability that they accru
when they dispose of hazardous waste.

The strict liability of Superfund has hit most major companies in this country through our enforcement efforts.

The strict liability of Superfund has hit most majo
companies in this country through our enforcemen
efforts. As a result, industries are very concerned wit
what happens to their waste, and they're lookin
harder and harder at how they can reduce their wast
in the first place. Even in the short-term, it's going t
make economic sense for them to analyze a recyclin
alternative as opposed to a disposal alternative. An
when they take into account the long-term economi
impact through long-term liability, the balance wil

clearly shift toward waste reduction alternatives.

JA: Given that waste reduction is a more attractive economic option than other waste management strategies, would a more structured regulation and enforcement of RCRA be a stimulus for greater economic development? [As it stands now, Section 3002 of RCRA requires a manifest and biennial reports from each toxic generator certifying that the volume of its hazardous waste is being reduced. However, the content of the certification and the nature of this waste reduction effort are left entirely up to the generator, with no civil or criminal penalties applying.]

LT: I remember sitting before the Senate committee when they were talking about that particular provision of the [RCRA] law, and I think the clear intent of Congress was: We want to acknowledge the goal of waste reduction as a part of this overall statute [RCRA] and we want to have industry move forward with that, but we want to do it in a different way than the traditional, regulatory way. We want to really try to promote a voluntary acceptance of this goal.

And that's how we've implemented it. Frankly, that's probably the best way to do it. The only way to get industry to really accept and incorporate waste reduction as a concept is to show them the economic advantages of it and to force them to think about it. We're not going to be able to structure it for them, or force them to do things. But we can force them to acknowledge the goal of waste reduction, to focus on that goal, and to look at the advantages it will give them.

The problem is that there's a vast array of industrial processes out there, each one industry specific. One of the real issues in waste reduction that you have to confront, and the reason why I think a lot of it has to be industries themselves, is that you're dealing with

processes that they don't want to make public to thei
competitors. How much they're willing to share is on
of the things that will limit how much your case studie
can tell.

I think you can talk generically about examples
even talk specifically in some cases, such as 3M. W
have to get industries themselves to really focus o
their own processes and deal with them. And if yo
show industry the economic advantages of waste re
duction, it *will* deal with them.

JA: So you intend to convince industries that waste reduc
tion is good for them?

LT: Yes.

JA: But some of the industry leaders that I've talked to sai
that the RCRA rules are very vague, and that there's
lot of confusion at the level at which the rules ar
written. They actually need a more specific, structure
regulation.

LT: Well, that's one of the things we'll work through as w
look at the report to Congress and our recommenda
tions as to whether any clarification of the regulation
are needed. I've heard some of that myself. But th
other side to the issue is that the more you push on
regulation, often the more strict that regulation be
comes and the more structured it becomes. I'm sur
that if we're not careful, we'll take away some of th
advantages of the [RCRA] regulation.

The key thing we want industry to do is to focus on what the overall objective of waste reduction is: environmental protection for us and economic payoff for them.

To me, the key thing we want them to do is to focu
on what the overall objective of waste reduction is: en

vironmental protection for us and economic payoff for them. How they get from point A to point Z down at the end is something that they've really got to work through. It's a fact that we want them to focus on Z, to develop a process for looking at the various alternatives and getting an answer. But we're not trying to precondition what the answer's going to be when they get there.

JA: Let me read you a list of suggestions from various industry leaders on how the EPA could help them to implement waste reduction strategies:

- A loan or grant fund from which industry can draw the monetary resources to finance waste reduction projects. The EPA-SBA [Small Business Administration] coordinated loan fund was supposedly discontinued in November 1985 because not enough people were using it.
- Provide funding to trade associations and universities with ties to industry. Many industry leaders cited both as being helpful in the education and implementation of waste reduction technology.
- Require the government and all government-funded contractors to employ only those companies that have an EPA approval stating that they are in compliance with RCRA.
- Reduce Superfund taxes for companies that are utilizing waste reduction.
- Provide a variance procedure in which EPA regulations can be relaxed for companies engaged in waste reduction.

Do you see any of these suggestions as viable, and if so, how do you plan to implement them?

LT: All of those are viable alternatives to look at. What you're basically asking is, 'What can government do that's an incentive, as well as what can government do to remove a disincentive?' The waste-end tax that we proposed for Superfund is an incentive tax for waste reduction. If a company minimizes waste, its tax is going to go down.

Our objective in proposing that tax was not only to raise revenue for Superfund, but also to be compatible with our overall goals under our regulatory program. We came up with a waste-end tax as one of the things that would be one more economic impact on industry to force it to look at waste reduction.

JA: Are you going to reduce those taxes for companies engaged in waste reduction?

LT: Well, the taxes would be reduced simply because the company's waste is reduced. They also would be higher for land disposal than they would be for treatment. So if a company has moved to a treatment approach, say incineration, which is the ultimate in waste reduction once you've gotten to a disposal stage, then there's going to be less waste for the government to tax.

A technical assistance approach is something we're going to try to promote as well, which is an information-sharing, clearing house approach. This seems to me to have a lot of potential. We will actually try to serve as a focal point to help bring information in from industry and get it back out to industry, as well as try to focus on specific industry segments, like the printed circuit board industry or the electroplating industry. We're looking at some possible alternatives within each industrial process to minimize waste, and then we'll put out a technical manual that industry can use.

JA: Your waste-end tax has actually gotten a lot of criticism, as far as its reliability is concerned.

LT: Yes, we've worked the reliability issue on the waste-end tax between our people and the treasury. Actually, it's the treater, storer, and disposal people who'll be the ones largely taxed. It's not the generator—he'll pay indirectly through an increased bill. We're at the point now where we've got the regulatory hold because the permits are coming from us; but the reliability issue stemmed from the tax collection problems that the Internal Revenue Service was having.

The problem is that there's just too vast a universe to have a reliable tax. We did a lot of sensitivity analysis on what effect waste reduction and treatment options would have on the total amount of revenue raised for Superfund. We projected anywhere from 300 to 600 million dollars a year as a realistic figure.

JA: Would any of that money be used to fund, say, technology development?

LT: No it wouldn't. Well, indirectly it would in that one component of the new Superfund bill is research and development, and a part of that research and development component, in the House and Senate versions, is to have a technology development program. But that program isn't aimed at waste reduction, it's aimed at new disposal technology, which is a different approach than you're aiming for. The tax is viewed primarily as a revenue source and secondarily as a waste reduction incentive.

JA: Do you think RCRA will make Superfund unnecessary?

LT: Oh yes, eventually it will. But you're going to see those two programs work together for at least the next ten years. RCRA deals with a fairly small universe of facility operators, as opposed to Superfund, which affects any facility out there that has a problem with a hazardous substance.

The Superfund program has a lot of clean-up work to be done over the next ten years, and I use ten years

The Superfund program has a lot of clean-up work to be done over the next ten years, and I use ten years as a time frame that I think we'll have most of the major problems behind us.

as a time frame that I think we'll have most of the major problems behind us. RCRA is going to be an intensive program, particularly on the treaters, storers, and disposers, and will probably become more intensive on the generators over the next few years.

JA: Getting back to our suggestions from industry, how about any funding, like the EPA-SBA loan fund?

LT: I don't have enough information on that program to be very responsive to you. I think funding for some special programs, like we've done with the League of Women Voters or the Tufts University program, are the kind of things that we'll continue to do. But that's more directed to the technical assistance, training kind of approach as opposed to a direct, grant program to industry. However, some of the programs that we've funded to states over the last couple of years under our state-grant programs have resulted in innovative, cost-sharing schemes with industry.

One thing we're doing right now is very controversial: For the disposal of Superfund waste, we will use only those facilities that are in compliance with RCRA, and those who go beyond RCRA.

JA: How about blacklisting? A number of industry leaders complained that many industries and government contractors use noncomplying companies because they're cheaper in their raw material supplies.

LT: I don't know exactly how that would work. One thing we are doing right now though is very controversial. It's not blacklisting, but some have called it that. For the disposal of Superfund waste, we will use only those facilities that are in compliance with RCRA, and those who go even beyond RCRA. If we're going to dispose of Superfund waste, then the company that's treating it has to meet very stringent requirements under our offsite disposal policy. We're actually requiring a more intensive review than RCRA does. But we haven't looked at whether we want to go beyond that in the sense I think you're talking about.

JA: What about a variance procedure in which EPA regulations can be relaxed for companies using waste reduction? Often the waste treatment system can produce initial violations until its methods are refined.

LT: That's one of the things we're looking at in our report to Congress, whether there are opportunities in the way you ensure compliance at a facility in order to promote waste reduction. As to exactly how that would work, and how it could be incorporated into a regulatory scheme, is one of the things we're looking at now.

JA: How about encouraging a federal tax-free status for bonds that could be issued to promote waste reduction? How about absolving generators from legal liability if they have a plan to detoxify or recycle their hazardous waste?

LT: The tax-free bond issue gets into a whole series of tax policy issues that I'm not very competent to talk about. As far as absolving from liability, there is some merit to really taking a hard look at that issue, as it relates to waste reduction.

Under Superfund, we will release from future liability any company that uses the best available treatment technology to clean up its hazardous waste site . . .

Under Superfund, and this would have some relevance to RCRA as well, we will release from future liability any company that uses the best available treatment technology to clean up its hazardous waste site, even if the waste stream at some future point in time presents an environmental problem. Then in fact what happens is that the government assumes the liability. But before that's done, we require the best available treatment alternative to be used, and in most cases that ends up being incineration.

JA: Can fines collected for RCRA infractions be used to promote or fund source reduction strategies for hazardous waste?

LT: Probably not. You're getting into the whole issue of fines coming into the Treasury as opposed to coming into an individual agency. And when I say 'probably not,' it's because you're confronting a far bigger issue than just RCRA fines; you're confronting the national issue of fines as a part of revenue for the federal government. What I *am* looking at is whether we can establish a fee system for some of our permitting programs, where we have those fees come into the agency to support innovative programs.

JA: How often is RCRA compliance reviewed for each independent generator, and how often are fully authorized state programs reviewed for enforcement?

LT: Well, fully authorized state programs are generally reviewed twice a year. Before a state's authorized, by the way, there's an in-depth review of that state, both of its regulations as well as its program.

With generators, because of the large volume involved, we will generally only review 10 to 15 percent of them a year for compliance. Because our enforcement resources are limited, our heaviest emphasis is on the disposers. We rely a lot on self-certification for the generator's part in compliance.

2

What Is Waste Reduction?

Although the public may not realize it, many American industries have progressively been putting less hazardous waste into the environment during the past five years. Studies conducted by the Chemical Manufacturers Association, for example, show that the chemical industry reduced wastewater discharge by some 19 percent from 1981 to 1984 and cut use of landfills by 35 percent during the same period.

These incremental reductions are cause for optimism. Even more compelling, however, are those firms that have not been content with incrementalism—that have, instead, gone for quantum leaps in waste reduction.

Consider, for example, 3M's estimate that it has shrunk its potential hazardous waste discharge by 50 percent during the past ten years and plans to cut its current levels by another 50 percent during the next five. Monsanto Corp. projects a similar reduction, targeted at 10 percent per year for each of the next five years.

As planning guru James F. Bandrowski points out, business planners often find themselves trapped by "timid incrementalism"—gradual increases in revenues or small decreases in costs. The same trap applies to planning efforts aimed at reducing hazardous waste. Although Lee Thomas speaks with guarded optimism about a "new era" in industry's approach to waste management, innovative approaches and stellar leaps toward the elimination of wastes have become a way of life in many firms.

The Institute for Local Self-Reliance has a similar goal: To make American cities independent, "extracting the maximum value from their raw materials, technologies, buildings, and their people." Like the cities, companies that reuse their waste will make themselves less dependent on "outside influences": suppliers of raw materials, independent waste handlers, the explosion in insurance rates, and the whims of public opinion, fueled by journalists hot for a new story on how industry tampers with the environment.

This, then, is the goal of waste reduction: to reduce as much as possible the waste being produced, or to recycle/reuse the materials as resources. In compiling the case studies of companies utilizing "waste reduction," we have taken a strict definition of the term. We include only those companies that have dramatically reduced (in some cases, eliminated) hazardous waste from their product manufacturing processes or have developed ways to reuse their previously wasted resources. We also include examples of nonhazardous waste reduction, where these examples involve especially innovative solutions.

There is much documentation out now of new technologies to detoxify hazardous waste, including chemical-eating bacteria, high-energy incinerators, and indestructible containers. **We do not recommend investing in such methods until all other waste reduction approaches have been used to their fullest extent.** Detoxification is a second-best solution because it still produces a waste stream, admittedly less harmful, but a waste stream nonetheless—one that requires further management. We are looking for long-term solutions, not quick-fix remedies. Often the waste reduction solutions require considerable capital investment. As the case studies in the following chapters point out, however, the investment can pay for itself in a relatively short period of time (usually less than two years; sometimes in only a few weeks).

$300 A DRUM—AND YOU'RE STILL LIABLE

We have all heard one horror story or another about waste disposal. Stories about companies spending in excess of $3 million to clean up an illegal site are not atypical. California Creative Dynamics Inc., for example, was caught illegally dumping hazardous waste into a sewer;

the president and vice-president were fined $10,000 and $5,000, respectively, put on three-year probation, and ordered to perform 500 hours of volunteer work. In addition, the company had to put an advertisement in an industry trade journal telling of their illegal dumping and warning others about California's tough, new law.

But even for legitimate permitted disposal, a company is not free from liability; that "safe" landfill may suddenly turn into a Superfund site, and when it does, the generators for the waste become liable, not the operators of the site. Companies are paying more and more, up to $300 a drum, for this liability. The 1984 RCRA amendments plan to phase out landfilling; according to one section of the law, by May 8, 1988, "no metalplater should plan to put his untreated metal hydroxide sludge in a hazardous waste landfill."

Throughout our case studies, the same reasons came up again and again as to why companies are implementing waste reduction options: the new, strict pollution regulations being enforced by the states, the rising cost of landfilling, or the unavailability of such disposal. However, nearly all these companies later realized other, even more important benefits than simply avoided disposal costs. Many found that the raw material they substituted turned out to be far cheaper, or that their new equipment was more energy efficient, or that product quality improved. Thousands, even millions, of dollars are being saved, not just in disposal costs, but in expenditures on energy, fuel, water, and raw materials. In other words, waste reduction is sound *economically* as well as ecologically. It's simply *good business.*

STARTING A WASTE REDUCTION PROGRAM

Let's say that you're convinced. You'd like to start your own waste reduction program. How do you go about it? A careful look at what other people have learned (see the case studies in the next chapters) suggests that a successful program involves several steps:

1. ***Make the commitment to waste reduction.*** If a waste reduction program is to get off the ground and involve the entire company, it must first have the support of top management, including the CEO or

president of the company. This means that management must educate itself on the subject, either by attending seminars sponsored by universities or trade associations, or by reading related articles and books. As Harry DeSoi, president of Pioneer Metal Finishing, says, "You have to stop being an industrialist and start becoming an environmentalist." Sometimes industries can help each other out in this education effort. For example, George McRae, plating manager for Stanadyne, Inc. in Sanford, North Carolina, started the "Lee County Waste Managers Association" in 1982. This association is open to representatives from all industries. At its meeting, the members share ideas and information on waste reduction and sometimes arrange for coordinated waste shipments in order to lower their unit transportation costs.

2. ***Communicate the program to the rest of the company.*** Middle managers and their employees are the company's greatest resource when it comes to new ideas for changing a product or process. Employees with their hands on the machinery are after all the ones closest to the manufacturing process and therefore the ones most likely to see where improvements can be made. The Minnesota Mining and Manufacturing Company (3M) used a 12-minute videotape and an 8-page brochure to communicate its "Pollution Prevention Pays" (3P) program to *all* employees. The videotape and brochure did two things:

- It displayed upper management's support for the program.
- It explained the program's goals: to make the company more cost-efficient as well as to improve the environment.

Many companies, however, have found that something else is needed before employees start thinking seriously about waste reduction. Usually this "something else" is a monetary incentive. A USS Chemicals plant in Ironton, Ohio, implemented a "Suggestions for Cost Reduction" program (SCORE) in July of 1981. The program rewards employees for their waste reduction ideas with a percentage of the money saved by implementing their suggestions. USS' employees have received a total of $70,000 for successful suggestions, which have saved the company $500,000 in costs. The Emerson Electric Company in Murphy, North Carolina, offers more modest incentives, such as color televisions, radios, and stereos. The incentives go to employees who suggest practical waste reduction ideas as well as to those who submit ideas for a cost-savings or a new product development.

3. ***Perform a waste audit.*** A company must know its waste before it can develop ways of not producing it. It must identify the hazardous chemicals in its products or in its waste streams, as well as the processes or equipment that are generating them. In the chapters that follow, we provide a list of hazardous chemicals and waste-producing processes that are common to the industries discussed, and in the opposite column, we provide a corresponding list of the waste-reduction solutions that have proved successful in reducing these wastes. A waste audit can vary in complexity and sophistication, from a fully computerized system such as is used by American Cyanamid Company, to a simple manual sampling and walk-through of the operations, such as would be used by smaller businesses.

Engineering interns from the local state universities have proved very helpful to a number of companies starting waste reduction programs. Peter Bohlman, human resources manager for Nordic Ware in Minneapolis, Minnesota, told how his company used the services of an intern from the University of Minnesota: "The intern taught us a lot about ourselves. In our situation, it wasn't a technical problem, it was a problem of time. The general manager was too busy doing his job to worry about reducing waste, whereas the whole reason for the intern's living and breathing was waste reduction." The Minnesota Technical Assistance Program, sponsored by the University of Minnesota, does a number of other things besides provide interns: it fields questions over the phone, sponsors seminars, and provides research grants to companies seeking to develop or improve upon waste reduction technologies (see chapter 8).

4. ***Analyze the costs and benefits.*** Modifications, especially if they involve the purchase of equipment, will have to be reviewed, either by a committee set up for that purpose, as in the case of large companies, or by the CEO or president of the company. In order for the committee or CEO to approve such changes, they have to know how much capital investment will be involved, and whether the company has the monetary resources for this investment. As Phil Horelick, president of Allied Metal Finishing, says, "If you want to do it right, you have to spend money." Here is where some waste reduction ideas, unfortu-

nately, end. This often happens when the analysis fails to integrate such factors as cost avoidance, enhanced productivity, and decreased liability risks into the study. It is important to remember that many waste reduction opportunities are low cost, have a quick payback, or both.

A number of state and local governments now provide grants and low-interest, long-term loans to companies interested in starting a waste reduction program. North Carolina's "Pollution Prevention Pays" technical assistance program, for example, is geared especially to small and medium-sized companies. Using money set aside by the state and provided by the Environmental Protection Agency (EPA), the program's committee awards "matching" grants of up to $5,000 to companies on the basis of need and the waste reduction project proposed. The recipient must then "match" the grant with a comparable investment, either with employee time or money. The state also has a revolving loan fund that lends waste reduction capital at low interest rates (4 percent). Since many companies have found that the payback on their waste reduction programs is less than a year, the loans are relatively easy to pay off. Minnesota offers a similar program. Other states now have laws that allow businesses to take accelerated write-offs on their taxes if they've invested in waste reduction (see chapter 8).

In 1977, the EPA, in conjunction with the Small Business Administration (SBA), created an $80 million fund that provided low-interest, long-term loans to companies needing financial assistance with their waste reduction programs but had been turned down by banks. Ironically, the EPA-SBA fund was discontinued as of November, 1985, because not enough businesses were using it. (Perhaps not enough industry leaders were aware of its existence.)

5. ***Implementing the program.*** Even after the budget obstacles have been cleared, resistance to the waste reduction project may still exist, simply out of distrust for anything "new." A manufacturing director, for example, may have solved a water pollution problem with a "black-box treatment" at the end of the pipeline. Then someone comes along with a plan to eliminate the source of the pollutant by changing the process or product. The director must decide between something that is familiar and works and something that should work better but is as yet unproven. Similarly, a sales manager may be concerned about customer acceptance of the changed product, which in the past has been performing well for customers.

When Harry DeSoi, president of Pioneer Metal Finishing, Inc. in Franklinville, New Jersey, decided to start a waste reduction program, he fired all employees who refused to become educated or participate in the program. There are less drastic ways, of course, to overcome such opposition. One method is to put such opposing persons on the planning and implementation committee; over the course of the planning and implementing process, these opposers often become so drawn into the program that they become its staunchest supporters.

6. ***Follow-up.*** Once the modification is in place, a company should monitor the extent to which pollution is being reduced and how much money is being saved in terms of energy, raw materials, and disposal fees. The information is not merely for interest's sake, but will be essential when the company files its manifest and bienniel reports on its waste reduction efforts under the 1984 RCRA amendments.

No matter how successful, a company's first waste reduction effort can always use improvements and new ideas, which the employees can provide if they are truly involved in the project.

THE OPTIONS AVAILABLE

There are many technologies and techniques involved in waste reduction. The following is a general, descriptive list of the waste reduction techniques most commonly used by various industries:

1. ***Housekeeping.*** Although it is one of the more obvious, commonsense approaches toward waste reduction, improved housekeeping is often the most effective first step for smaller companies.

The Emerson Electric Company's waste audit, for example, identified a number of problems, most of them simply matters of "poor housekeeping." Storing hazardous waste outside the facility and improperly labeling it increased the risk of spillage, which would contaminate the surrounding surface water. Unauthorized dumping of chemicals into the on-site water treatment system overloaded its treatment capacity, thus violating effluent guidelines. Correcting these problems eliminated the company's violations of the environmental regulations and reduced its risk exposure. Educating employees on the proper disposition of chemical waste makes for a safer work environment as well.

Stanadyne, Inc., a manufacturer of plumbing products, realized that plating a defective part created just as much waste as plating a good one, and thus it began to pre-inspect parts. ICI Americas, Inc., a chemical research and development firm, decided to separate its wastes according to type, either as chlorinated or nonchlorinated solvents. This separation allowed for the recycling of the solvents back into their separate processes.

2. ***Material Substitution.*** This solution, which involves replacing the polluting ingredient of a product with an less toxic one, may be the most logical and complete way to eliminate waste. However, the material substitution approach may be applicable only in certain situations. It is at its most economical when a product is being developed for the first time, as in the case of the Minnesota Mining and Manufacturing Company when it developed a new fire extinguishing agent. 3M's first product was very effective, but the chemicals were known to cause serious environmental problems. At the laboratory, researchers isolated the two harmful ingredients and found substitutes. The final product has 1/40th the adverse effect of the original, and in addition, is far cheaper to produce.

With an existing product, however, the substitution approach can be expensive, since the new chemicals often require an additional investment in new equipment. The alternatives, however, can be even more expensive.

Riker Laboratories of California, for example, previously used an organic-based solvent as a carrier for coating its pharmaceutical tablets. Faced with stricter air pollution regulations, the firm first investigated the possibility of installing pollution control equipment: The initial cost of such equipment would've been $180,000, and in addition, $30,000 would've been required to operate the equipment for an entire year. As an alternative to such a high expense, Riker decided to explore the waste reduction approach. By making a $60,000 investment in equipment modifications, the company was able to switch from an organic solvent-based coating to a water-based coating. This eliminated the pollution problem completely, obviating the need for pollution control equipment. Moreover, the water-based material worked just as well in the tablet coating process—and resulted in a $15,000 per year savings in raw material costs.

3. ***Equipment Redesign.*** Sometimes it's not the product that causes the pollution, but the process used in its manufacture. The solution is similar to material substitution: Replace the old equipment with something else that doesn't produce the harmful byproduct.

Pioneer Metal Finishing, Inc. invested $210,000 in facilities and equipment for their new "batch" recycling system, and the savings that the system produced paid back their investment in three years. President Harry DeSoi was able to finance the project with a long-term, low-interest federal loan.

4. ***Recycling and Reuse.*** The array of recycling and reuse options is far too extensive for detailed discussion here. Which one a company decides to use depends on the chemical makeup of its byproducts. Many industries have been able to invest in on-site recycling equipment, and thus have solved or reduced both their waste management

Exhibit 1: Distillation. In automatic distillation units, the sludge is pumped away as it is deposited. In manual systems, the still is shut down periodically to manually scrape off and remove the sludge.

Source: Brighton Corporation, Cincinnati, Ohio, 1982.

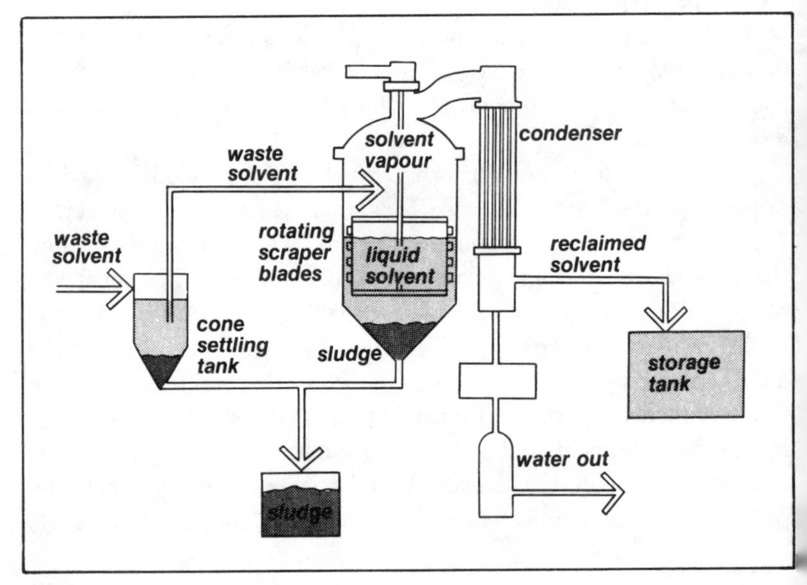

cost problems and their raw material needs. For the smaller industries that can't afford to invest, there are numerous off-site recyclers who will take the wastes, recycle them, and sell them back to the company at a much lower price than for virgin material. For those interested in developing on-site recovery facilities, the recovery processes include:

a) *Distillation*—mainly used to recover organic solvents. A unit distills the solvent and recondenses the solvent vapor for reuse (see Exhibit 1). A typical distillation unit recovers 85 to 90 percent of the original solvent.
b) *Absorption*—in this process, a bed of carbon or other compound is used to absorb gaseous materials or evaporating solvent vapors (see Exhibit 2). The compound can be recovered in a highly concentrated form when the absorbent beds are regenerated, usually with the application of steam. Absorption units can achieve 90 percent solvent recovery.

Exhibit 2: Absorption. Solvent-laden air is passed through a bed of activated carbon that retains the solvent. In the desorption phase, solvent is stripped from the carbon bed and condensed to a liquid.
Source: "Disposal of Surface Coating Solvents," *Industrial Finishing*, February 1981.

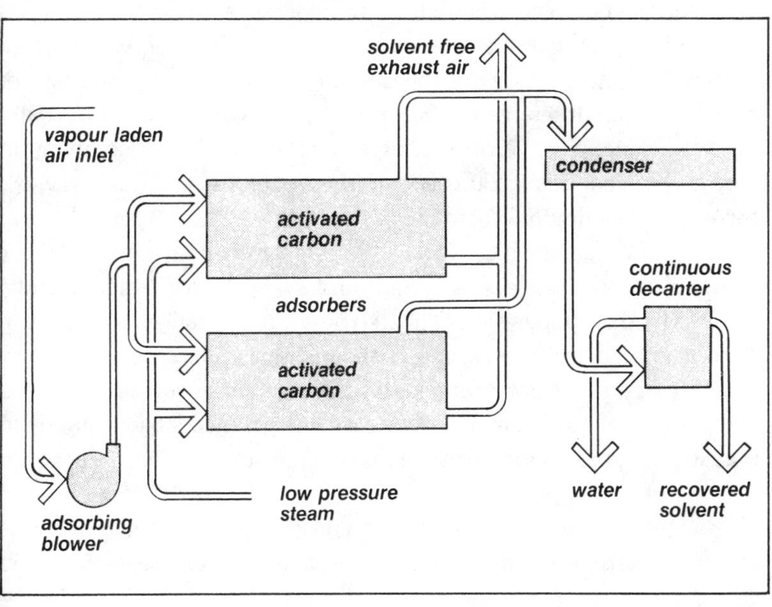

c) *Filtration*—waste streams containing particulate matter can be passed through filters, which capture the particles for reuse. The porosity of the filter will determine which particles are recovered. Some industries add lime to their waste stream before passing it through the filter; in this way heavy metals are precipitated and more easily removed. Reverse osmosis as well as other techniques are proving useful. See the box on page 43.

d) *Electrolysis*—an electrically charged cell attracts ions and removes them from the waste stream. The electroplated metal can then be removed from the cathode and sold as scrap. Many metals can be removed in this way, including copper, manganese, nickel, chromium, and silver. The recovery rate is excellent, ranging from 90 to 99.9 percent.

7. ***Waste Exchange.*** Materials that one industry views as waste may actually be a valuable resource in another industry. Informational networks called "waste exchanges" operate like the "personals" of popular newspapers: Industrial firms fill out forms describing the waste they produce or the raw materials they need. This information is then printed in a publication such as the "Waste Watcher" (see Exhibit 3 on opposite page), which is distributed to all industries on the mailing list. The search is complete when a firm is contacted by someone that needs and is able to recycle the raw materials present in their waste. Both parties benefit, either from the sale and avoided disposal costs of previously unmarketable waste, or from receiving precious raw materials at an inexpensive price. For example, a chemical company in North Carolina sells 5,000 gallons of hydrochloric acid waste a week through the Piedmont Waste Exchange, earning $70,000 per year from the sale. Another North Carolina company has been saving $40,000 per year in disposal costs ever since it found a recycler for its spent acids.

This kind of cooperation exhibited by waste exchanges is essential if American industry is to survive in the national and international marketplace. Consider, for example, that U.S. industry currently imports 100 percent of its chromium. Waste exchange, by enabling American industry to reuse its raw materials, will make our country less dependent on foreign suppliers.

8. ***Detoxification.*** Detoxification processes neutralize hazardous waste into harmless materials. However, as we said before, they are the

Exhibit 3. Materials listing form that customers must fill out in order to be listed in the *Waste Watcher*, the quarterly bulletin of the Piedmont Waste Exchange.
Source: Piedmont Waste Exchange, Charlotte, North Carolina, 1985.

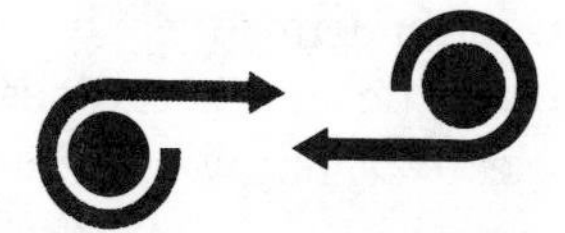

PIEDMONT WASTE EXCHANGE

Urban Institute
University of North Carolina at Charlotte
Charlotte, N. C. 28223

MATERIALS LISTING FORM

1. Company Name: ______ SIC Code: ___ ___ ___ ___
2. Mailing Address: ______
3. Company Contact: ______ 4. Title: ______
5. Signature: ______ 6. Date: ______ 7. Phone ( ) ______
8. Check One Only: ☐ MATERIAL AVAILABLE ☐ MATERIAL WANTED
9. Classifications (review all first, then select one only that best describes your material):
   ☐ ACIDS ☐ ALKALIS ☐ OTHER INORGANIC CHEMICALS ☐ SOLVENTS ☐ OTHER ORGANIC CHEMICALS ☐ OILS AND WAXES ☐ PLASTICS AND RUBBER ☐ TEXTILES AND LEATHER ☐ WOOD AND PAPER ☐ METALS AND METAL SLUDGES ☐ MISCELLANEOUS
10. Material to be listed (main usable constituent, generic name): ______
11. The industrial process that generates this waste: ______
12. Main constituent (chemical formula): ______ % ______ in ______
13. Contaminants (highest to lowest, chemical formula): ___ % ___ ___ % ___ ___ % ___ ___ % ___
14. Percentage by (check one): ☐ VOLUME ☐ WET WEIGHT ☐ DRY WEIGHT
15. Physical State: ☐ SOL'N ☐ AGGREGATE ☐ SLURRY ☐ SOLID ☐ SLUDGE ☐ DUST ☐ CAKE ☐ GAS
16. Miscellaneous appropriate information (pH, toxicity, reactivity, color, particle size, flash point, total solids, purchase date): ______
17. Potential or intended use: ______
18. Packaging: ☐ BULK ☐ DRUMS ☐ PALLETS ☐ BALES ☐ OTHER ______
19. Present amount: ______ 20. Frequency: ☐ CONTINUOUS ☐ VARIABLE ☐ ONE TIME
21. Quantity thereafter: ______ in ☐ lbs. ☐ Litres ☐ tons ☐ Cubic Meters ☐ gal. ☐ C.y. ☐ Other ______ ☐ Kg. ☐ Tonnes
    per: ☐ Day ☐ Week ☐ Month ☐ Quarter ☐ Year
22. Restrictions on amounts: ☐ none ☐ minimum ☐ maximum ______
23. Available to Interested Parties: ☐ sample ☐ lab analysis ☐ independent analysis
24. For material wanted, acceptible geographic range (i.e.,states, provinces, regions, countries): ______
25. For material available, if location of the material differs from the above mailing address, complete the following:
    City ______ State or Province ______ Area Code ______
26. This listing is ☐ CONFIDENTIAL ☐ NON-CONFIDENTIAL (To expedite inquiries, the Exchange may give my name and telephone number to inquirers.)

LISTING FEE:
$40 per company (unlimited listings) for four consecutive issues (published quarterly).

Payment must be made in advance.

Please include your check for $40 made payable to Piedmont Waste Exchange.

least desirable solutions because waste still remains after the detoxification process has been completed. Thus, detoxification should be considered only after the previous seven approaches have been tried. Detoxification can include any of the following approaches:

1. Chemical neutralizations: An acidic waste, for example, can be used to neutralize a basic waste, thereby rendering them both safe. Or a chemical can be used to break down a hazardous compound, as is done with cyanide in electroplating waste solutions.
2. Biological (microbiological): Treatment with microorganisms can break down many organic compounds in some hazardous wastes.
3. Thermal: Various thermal detoxification methods, including incineration and wet oxidation, are now available for detoxifying organic compounds.

Incineration is useful because, in addition to destroying the waste, it also generates heat energy that can be put to a number of uses. United Globe Corporation, a furniture manufacturer in Lexington, North Carolina, installed a two-stage incineration system that not only detoxifies semisolid and liquid finishing wastes, but also produces steam that the company uses for heat and in its ragwash operations. United Globe reports savings of approximately $905,000 a year in avoided land disposal costs and $36,000 a year in energy costs.

A company must ensure, however, that its toxic substances will be completely destroyed in the incineration process. Polychlorinated biphenyls (PCBs) and dioxin, for example, are only broken down when burned at very high temperatures (exceeding 2,400 degrees Fahrenheit). If the incinerator is not capable of generating such heat, the company might become liable for releasing toxic gases into the air.

The following chapters address waste reduction as it relates to specific industries. Those industries include: Chemicals and Allied Products (standard industrial code [SIC] 28); Fabricated Metal Products (Electroplating, SIC 34); Nonelectric and Electric Machinery Manufacture (SICs 35 & 36); Electronic Equipment (Printed Circuit Boards, SIC 36); and Electric, Gas, and Water Utilities (SIC 49). Each chapter begins with a list of the waste-producing products or processes on the left-

SCREENS, FILTERS, AND BUBBLES

A number of physical processes can be used to separate valuable components from mixed streams. Each process has its specific applications, its strengths and weaknesses, and each is currently the subject of ongoing research. Although space limitations prevent a detailed discussion, an overview is in order.

1. Electrodialysis uses an applied electrical field to draw ions through a selectively permeable membrane. These physically rugged membranes are often composed of a polystyrene matrix crosslinked with divinylbenzene and rendered either anion or cation selective.

2. Reverse osmosis (RO) separates macromolecular or ionic components from a solution by using a semipermeable membrane. Pressure reverses the normal osmotic flow of water across the membrane, thus concentrating and purifying the solution.

3. Ultrafiltration separates colloids or macromolecules from lower molecular weight solutes. Membrane components include cellulose nitrate, cellulose acetate, polysulfone, aramids, polyvinylidene fluoride, and acrylonitrile polymers.

4. Liquid membranes offer a promising new approach to solving the problems of polymeric membranes. A liquid membrane is formed by the emulsion of two immiscible phases and dispersion of that emulsion into a third (continuous, or donor) phase.

5. Absorption is the most widely used non-vapor-liquid technique for molecular separation in the petroleum, petrochemical, and chemical industries. A microphorus solid (such as activated carbon, molecular sieve zeoline, silica gel, or alumina) captures components from a liquid or gas. These components can be recovered by increasing the temperature or reducing the pressure of the absorbent.

6. Bubble or foam separation is used in the large-scale removal of trace quantities or material from aqueous media. The trace quantities attach preferentially to the air/liquid interface of a bubble or foam, which then rises through the liquid.

hand side, and the relevant waste reduction solution(s) on the right. The chapter then goes on to discuss in-depth each waste and its solutions.

All kinds of companies, from small to large, have implemented these solutions, and specific examples are given throughout each chapter. In addition, the waste reduction solutions that work for one industry can often be applied to other industries using similar manufacturing processes. (For instance, tape manufacturers, medicine tablet manufacturers, and paint applicators all use a similar drying process for their products.) However, an individual company may have pollution problems that are idiomatic to itself, and therefore many of our solutions may not be applicable, or there may be solutions of which we have not even heard. At the end of each chapter, we provide the addresses of each company or name mentioned, in order of their appearance, so that our readers may contact them if they desire further information.

3

Chemicals and Allied Products (SIC 28)

1. EMISSIONS FROM ORGANIC SOLVENTS

Floating Roofs and Conservation Vents. Volatile organic solvents, such as acetone and methyl ethyl ketone, evaporate quickly, even in room temperature, and thus large amounts of solvents may be lost when stored under ordinary conditions. One solution, however, has proven effective in reducing volatile organic compound losses: floating roofs. These are simply covers resting on the surface of the solvent, rising and falling as the level of the liquid changes within the enclosing tank.

Exxon Chemical Americas' Bayway plant in Linden, New Jersey, installed 11 of these floating roofs in 1975, just as New Jersey was about to pass new air pollution control laws that would require floating roofs and conservation vents for the reduction of many sources of organic emissions. Although the original motive for installing these roofs was to comply with the new regulations, Exxon subsequently found that the roofs conserve a substantial amount of raw material, which in turn saves them money. Consequently, they installed five more roofs on new or existing tanks. All in all, five million pounds of volatile organic

Waste-Producing Product or Process	*Waste Reduction Solution(s)*
1. Emissions of volatile organic solvents, stored in holding tanks.	Floating roofs and conservation vents to contain solvent vapors.
2. Evaporating solvent from the drying of adhesive tape. Cumene emissions from phenol manufacture.	Absorption recovery using activated carbon or resin beds. Installation of condensor units. Nitrogen-based solvent recovery. Material substitution.
3. Organic solvents used in the manufacturing process or the cleaning of equipment.	Segregation of waste streams. On- or off-site distillation.
4. Organic solvents used to coat medicine tablets.	Material substitution and equipment redesign. Recovery using carbon absorption or nitrogen-based drying equipment.
5. Calcium fluoride precipitate from the manufacture of compounds.	Reuse of calcium fluoride in another manufacturing process.
6. Waste dust from pesticide formulation	Equipment redesign to segregate waste streams.
7. Rejected perfume, cologne, and toilet water products classified as hazardous.	Incineration

compounds have been saved since 1975, and in 1983 alone, 680,000 pounds were saved with a value of over $200,000. Capital investment for the roofs ranged anywhere from $5,000 to $13,000 per roof, depending on the size and complexity of the tank; even the most expensive installations, however, paid for themselves in one year.

Conservation vents, also required by New Jersey state law, were installed on seven of Exxon's storage tanks and process sources. These vents are used to control the emissions of such chemicals as phenol, maleic anhydride, and other hydrocarbons. Conservation vents release a tank's accumulating vapors less freely and continuously than the standard vents that are routinely installed; they can reduce air emissions by 30 to 75 percent. Detailed information on material savings, such as that available for the floating roofs, was not provided for the conservation vents.

2. EVAPORATING SOLVENTS/CUMENE EMISSIONS

Adhesive manufacturers use solvents, such as toluene, to dissolve the rubber and resin components of adhesives in tape products. During the drying process, the solvent evaporates, creating air pollution when exhausted from the plant. Carbon absorption equipment effectively captures the evaporating toluene, which is then returned to the process for reuse.

Similarly, cumene vapor is emitted during the manufacture of phenol, a chemical that is sold and used in other manufacturing processes. The price of cumene has risen steadily, from 5¢ a pound during the 1970s to 30¢ a pound in recent years; this rising cost has motivated many chemical companies to consider ways to recover the cumene previously lost in the manufacturing process. Carbon absorption has likewise proven useful in capturing the escaping cumene vapors, enabling their return to the manufacturing process.

Absorption Recovery. Absorption systems, which recover solvents from solvent-laden air, are well-known in the chemical industry; the technology has been around for at least 40 years. Companies, especially small ones, have been reluctant to use absorption because of the capital investment required; this may change as solvent prices continue to increase, and as RCRA forces companies to look at waste reduction technologies. Absorption systems consist of three stages:

a. A bed of activated carbon or a resin absorbs the solvent vapors from the passing air. Many new types of resins are now commercially available, and a specific resin can be purchased to selectively absorb certain materials.
b. The saturated carbon or resin is then stripped of the solvent temporarily attached to it; this is known as "desorption." Desorption can be accomplished by several methods: low-pressure steam; a solvent that removes the absorbate; acidic or caustic wash followed by thermal activation of the carbon at 1,000 degrees C; and indirect heating with a nitrogen purge. Steam is most frequently used because it is inexpensive, easily condensed back to a liquid, and doesn't present a disposal problem.
c. The solvent in solution is recovered, either through simple decantation if the solvent is immiscible in water, or otherwise through distillation, such as is described in section 3.

Shuford Mills, a tape manufacturer in Hickory, North Carolina, installed two carbon absorption systems, which achieve a 95 percent recovery of the toluene used in their adhesive coating operation. Although the systems were costly to install, Shuford Mills expects that the equipment will pay for itself in four years.

USS Chemicals' Haverhill plant in Ironton, Ohio, cited rising cumene costs and the availability of new types of resins as their main reasons for installing a resin absorption system. This system reduced cumene emissions by 80 percent, and is now saving 715,000 pounds of cumene each year, worth $175,000.

Condensor Recovery. USS Chemicals found another way to recover cumene from its phenol manufacturing process. A plant operator noticed that vapors were escaping from a pressure control vent in the manufacturing equipment. He suggested that the company install a surplus condenser to capture these emissions and return them directly to the manufacturing process in their liquid form. USS Chemicals carried out his suggestion in 1983. In its first year of operation, the condenser unit recovered 400,000 pounds of cumene worth $100,000; the unit cost only $5,000 to purchase and install.

Nitrogen-Based Solvent Recovery. Most conventional drying ovens need a constant source of air in order to dilute solvent concentra-

Exhibit 4: Nitrogen-Based Solvent Recovery. With an inert nitrogen atmosphere, solvent vapor can be safely concentrated to well above traditional oven levels, at which levels the solvent is recoverable by condensation. The oven exhaust is cooled with multiple heat-exchange stages. Solvent recovery reaches 99 percent efficiency.

Source: "Solvent Recovery System Saves Costs and Cleans Air," *Chemical Engineering*, March 10, 1980.

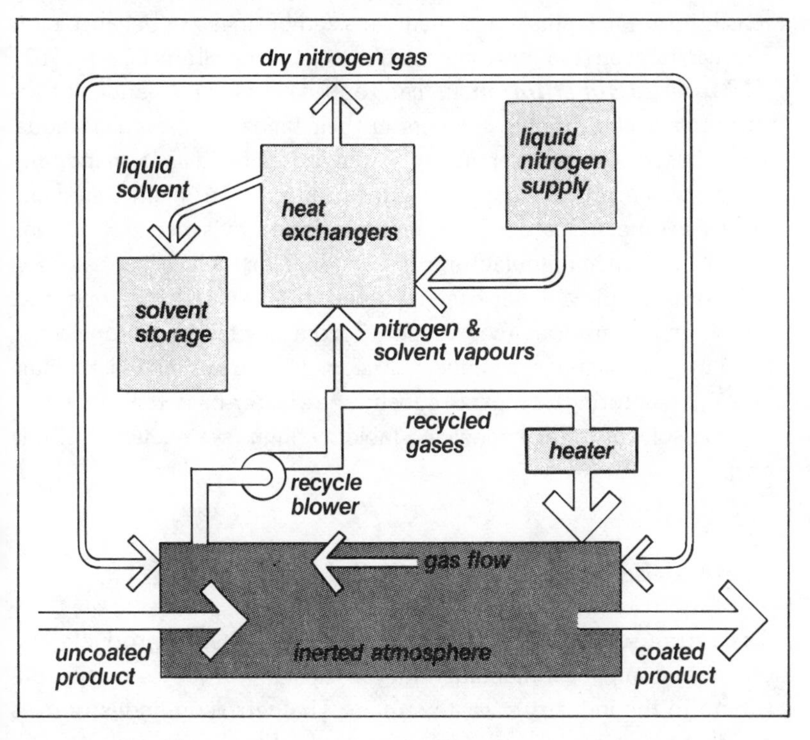

tions below explosive levels. The hot air continuously exhausted from the system carries with it a lot of energy. Airco Industrial Gases in Murray Hill, New Jersey, is marketing a new system that replaces the air with an inert nitrogen atmosphere (see Exhibit 4). The nitrogen allows the solvent vapor to exist at much higher concentrations than in conventional ovens, and the system is entirely self-contained, thereby eliminating heat loss and safely capturing the previously lost solvents.

Airco claims that the nitrogen-based system is more energy and recovery efficient. The equipment was tested at a fabric-coating facility in Kenyon, Rhode Island, and was found to reduce heating energy requirements by 60 percent over the previous system, saving $7.75 per hour. And because the solvent is at a much higher concentration, it can be recovered directly by condensation; this translated into a 99 percent recovery efficiency. However, nitrogen-based technology may not be feasible for some small- and medium-sized businesses, because of its high capital cost. (For more information, see Campbell and Glenn, 1982.)

Material Substitution. In many applications, tape manufacturers can replace the organic solvents in their tapes with nonhazardous, water-based solvents, depending on the product quality demands and the particular process used. Such a substitution has been tested and proven in other industries: The Hamilton Beach Division of Scovill, Inc., a small appliance manufacturer in Clinton, North Carolina, has been able to substitute a water-based cleaner for a solvent-based one in its degreasing operations (see section 1 from Electric and Nonelectric Machinery). The pharmaceutical division of 3M (the Minnesota Mining and Manufacturing Company) substituted a water-based solvent for an organic solvent in their medicine tablet coating (see section 4 of this chapter).

3. MANUFACTURING AND CLEANING SOLVENTS

Aside from solvent vapors, various liquid solvents, by-products of a chemical company's manufacturing or cleaning processes, also contribute to the industry's waste stream. The petroleum industry uses solvents to extract lube oils and waxes during the refining process. The pharmaceutical industry uses alcohol-based solvents, such as methanol and butyl alcohol, in their extraction processes. The plastics industry uses hexane, benzene, and cyclohexane in the manufacture of high-density polyethylene and polypropylene. Paint formulators and adhesive manufacturers use solvents to clean out their equipment. According to the authors of *Profit from Pollution Prevention*, a guide to reduction and recycling published by the Pollution Probe Foundation in 1982, there are currently a "few dozen" organic solvents recycled in industry; these include halogenated hydrocarbons, fluorocarbons, methylene chloride, perchloroethylene, trichloroethylene, 1,1,1-trichloroethane, and various alcohols, nonhalogenated ethers, keytones, and aliphatic compounds. Since 1982, solvent recycling has become more widely practiced, and thus many more solvents have been found to be recyclable.

Solvent Segregation. The first step in solvent recovery is for the company to segregate its waste stream in two ways:

a. Into chlorinated and nonchlorinated solvents. ICI Americas, Inc., a chemical research and development facility in Goldsboro, North Carolina, reduced its hazardous waste stream by 70 percent simply through segregation and subsequent distillation of these two classes of solvents.
b. Into individual solvent components. This may mean walking through the manufacturing process to see where waste solvents are currently combined, and then ensuring that individual solvents are kept separate. This separation will make recovery much easier and less expensive. Fractional distillation can be used to separate combined solvents into their individual components.

Distillation. Distillation separates a particular solvent from its contaminants by boiling it out of solution. The desired solvent is then recondensed in another container for reuse (see Exhibit 1). Many non-halogenated organic solvents are ideal for distillation because they usually have low boiling points, and a typical distillation unit recovers 85-90 percent of the original solvent.

As discussed in section 1 of Electric and Nonelectric Machinery, the most important decision a company must face when using distillation is whether to recover the solvent in-house or employ the services of an off-site recycler. Off-site recovery eliminates the capital investment and maintenance and operating costs needed for distillation equipment. However, the generator must assume liability for his waste during transport to the recycling facility, and sometimes the returned product is contaminated with unwanted substances from another's waste.

The American Enka Company, a nylon yarn production and research facility in Enka, North Carolina, found that distilling their isopropyl alcohol solvent in-house is more profitable than having the solvent recycled by an outside firm.

Previously, American Enka used an outside firm to recycle the alcohol, and they reported many problems with the arrangement. Distillation losses averaged 15 percent, but losses as high as 40 percent occurred. Sometimes the outside firm didn't clean the distillation column between runs, which meant that the returned alcohol was contaminated with Dowtherm, benzene, ethyl benzene, methyl benzene, and various chlorinated hydrocarbons. Thus, each returned batch had to be tested for contamination, and if found unusable, the batch had to be disposed of by incineration.

To solve this problem, American Enka purchased a used distillation unit for $7,500, and modified it to distill the isopropyl alcohol. American Enka is now saving $90,000 a year in avoided virgin solvent and outside distilling costs. Their in-house system is more efficient, recovering 90 percent of the solvent as opposed to the 85 percent efficiency of the outside firm. Even the still bottoms are recycled; they are used as an asphalt emulsifier in another product line. To sum up, American Enka reduced its hazardous waste production by 10,000 gallons per year, while the equipment investment paid for itself in only one month.

4. TABLET COATING SOLVENTS

Material Substitution/Equipment Redesign. In the pharmaceutical industry, medicine tablets are coated using organic solvent carriers. In some cases, these organic solvents can be replaced with water-based solvents, which eliminate air pollution, cut solvent costs, and obviate the need for air pollution control equipment. Different equipment may be required to handle the water-based solvent.

Riker Laboratories, a pharmaceutical division of 3M, decided to switch its medicine coatings from organic solvents to water-based solvents when the state of California imposed new, stricter air pollution limits. The substitution required different spray equipment to be installed in the coating machine, and the pumping mechanism, tubing, and control systems also had to be modified. Organic solvents have a higher volatility than water, which means that the heating requirements of the drying ovens will be less for oganic solvents than for water solvents. Consequently, the substitution required that the heating capacity of the dryer be increased so that the tablets dried properly. The total cost for the equipment modifications was $60,000.

No change in product quality occurred due to the substitution. In addition to preventing 25 tons of air pollution each year, the water-based solvent is saving the company $15,000 a year in solvent costs because it is far cheaper to purchase than its organic predecessor. Air pollution control equipment, which would have cost $180,000 to install and $30,000 a year to operate, was not needed, thus saving the firm substantial capital investment and operating costs.

Solvent Recovery. For some pharmaceutical companies, the switch from organic solvents to water solvents in the product coatings is impractical or compromising to product quality. For these companies, proven technology is available to recover the organic solvent which evaporates during the drying of the tablets. Since solvent coatings are dried in a similar manner to that used by adhesive manufacturers for their tapes, the carbon absorption techniques and nitrogen recovery technology discussed in section 2 of this chapter will apply here.

5. CALCIUM FLUORIDE PRECIPITATE

The manufacture of inorganic fluoride compounds, such as uranium hexafluoride and sulfur hexafluoride, typically produces waste streams containing large amounts of fluoride ions. These ions are usually precipitated with lime as calcium fluoride, which, according to the new RCRA regulations, is a hazardous waste because of its high pH.

Reuse of Calcium Fluoride. Instead of disposing of the calcium fluoride, a company would be wise to take a closer look at its waste. Depending on the treatment process used, the precipitate can be used in another manufacturing process, or sold to another industry that needs this raw material.

Allied Corporation's plant in Metropolis, Illinois, a nuclear fuel and chemical manufacturing facility, was forced to look for novel ways to handle their waste when no more surface impoundments could be constructed because of the scarcity of land. Allied examined its sludge and found that it's primary component was calcium fluoride, together with 10-15 percent unreacted calcium hydroxide and 10 percent sulfates and other fluorides. These other solids are now processed in a wastewater treatment plant, which reacts the solids with dilute hydrofluoric acid. The acid neutralizes the sludge and converts the remaining free lime to calcium fluoride.

Allied realized that its calcium fluoride precipitate could be used at another site in the manufacture of anhydrous hydrofluoric acid. Thus, 8,000 tons of waste fluoride each year are turned into a valuable raw material, saving the company $1 million a year. In avoided disposal and storage costs alone, $300,000 is being saved through the modification. The initial investment for recovery and treatment equipment was $4.5 million, but the equipment should pay for itself in 4 1/2 years.

6. WASTE PESTICIDE DUST

Equipment Redesign For Waste Stream Segregation. Sometimes an industry can reduce waste simply by ensuring that the individual waste streams are kept separate; once they are mixed, the material may be no longer usable or their recovery may be too expensive.

The Daly-Herring Company, a pesticide formulation facility in Kinston, North Carolina, generates approximately 45,000 pounds of pesticide dust each year from two major production systems. Previously, the dust from several processes was collected in a single baghouse and couldn't be reused because it contained a mixture of several pesticides. Over a hundred drums of waste dust were landfilled each year at a total cost of $3,200 per year.

In 1983, the firm removed its single baghouse collection system and in its place installed vacuum-air collection equipment for each production line. Dust from each line is now collected separately in independent baghouses; it is then filtered and automatically shaken down into a hopper which augers the dust back to the production line from which it came.

The returned dust is used as an inert filler in the final product, because the quantitative analytical testing required to characterize the dust as an active ingredient is too costly.

The recycling saves the company more than $2,000 in raw material costs, and eliminates waste dust disposal costs, saving an additional $9,000 per year. A $9,600 investment was required for new equipment and renovations, yet the equipment paid for itself in less than a year.

7. REJECTED AND OUTDATED COLOGNES, PERFUMES AND OTHER TOILETRIES

Incineration. In the toilet goods industry, a certain percentage of products inevitably becomes damaged or outdated; these products must be destroyed because they constitute a hazardous waste under RCRA. Usually, bottles containing rejected cologne, perfumes, and toilet waters are crushed, and the residual liquids and containers are sent to a hazardous waste landfill.

The Coty Division of Pfizer, Inc., a toilet goods manufacturer in Sanford, North Carolina, realized an alternative solution besides landfilling for their product waste. Since most toilet goods are hydro-alcoholic in nature, they have a BTU value of approximately two-thirds that of fuel oil. Coty now saves its rejected products and mixes them with fuel oil for use in their heating system.

This required minor modifications to the existing hot water boiler and the construction of a collection apparatus and a special feed system. The total cost for the project was $7,500, however, the company is saving $2,800 a year in waste disposal and fuel oil costs. For every gallon of hydro-alcoholic product consumed, two-thirds of a gallon of fuel oil is conserved. The new equipment paid for itself in less than three years, and no hazardous air pollutants are released from the boiler, nor are any hazardous wastes buried in landfills.

Contacts For Futher Information

1. Exxon Chemical Americas
   Research and Engineering
   1600 East Linden Avenue
   Linden, New Jersey 07036
   Contributing expertise: Floating roofs and conservation vents.

2. Shuford Mills, Inc.
   1506 Highland Avenue, Northeast
   Hickory, North Carolina 28601
   Contributing expertise: Carbon absorption recovery.
3. USS Chemicals
   P.O. Box 127
   Ironton, Ohio 45638
   Contributing expertise: Resin absorption and condensor recovery.
4. Airco Industrial Gases
   575 Mountain Avenue
   Murray Hill, New Jersey
   Contributing expertise: Nitrogen-based solvent recovery.
5. Paul Ekoniak, Senior Registration Specialist
   ICI Americas, Inc.
   P.O. Box 208
   Goldsboro, North Carolina 27533-0208
   Contributing expertise: Solvent segregation.
6. John Ray, Environmental Engineer
   American Enka Company
   Enka, North Carolina 28728
   Contributing expertise: Distillation recovery.
7. Mike Koeningsberger, Environmental Engineer
   Minnesota Mining and Manufacturing Company (3M)
   3M Center
   St. Paul, Minnesota 55101
   Contributing expertise: Material substitution of tablet coating solv
8. Edward Shields, Director of Environmental Services
   Allied Corporation
   Columbia Road and Park Avenue
   Morristown, New Jersey 07960
   Contributing expertise: Reuse of calcium fluoride precipitate.
9. D.W. Craig, Vice-President/General Manager
   Daly-Herring Company
   P.O. Box 428, Neuse Road
   Kinston, North Carolina 28501
   Contributing expertise: Equipment redesign to segregate waste dust streams.

10. R.R. Clarke
    The Coty Division of Pfizer, Inc.
    P.O. Box 1026
    Sanford, North Carolina 27330
    Contributing expertise: Incineration of rejected and outdated toiletry products.

4

Fabricated Metals (Electroplating) Products (SIC 34)

1. DRAGOUT FROM PLATING

The fabricated metal products industry manufactures ferrous and nonferrous metal products such as tin cans, tinware, hand tools, cutlery, general hardware, nonelectric heating apparatus, fabricated structural metal products, and metal forgings. The metal parts are plated, usually with copper, nickel, zinc, chrome, tin, or cadmium, in order to produce an attractive finished appearance and protect the metal from corrosion. The plating is accomplished either manually or with automatic machines, such as a Udylite Cyclemaster. In both cases, as parts are lifted out of the plating bath, the solution that clings to the part is called *dragout*, which becomes pollution in the rinsing process. Reduction of dragout is essential not only for waste reduction but also for conserving the plating bath and keeping it free from contamination.

Solution Maintenance. Before dragout reduction is attempted, platers should make sure that their plating solutions are properly main-

Waste-Producing Product or Process	*Waste Reduction Solution(s)*
1. Dragout from plating operations.	Solution maintenance using electrolytic recovery or filtration. Raw material conservation. Material substitution. Equipment redesign. Reduction of dragout using spray rinses.
2. Chromic acid and copper ions from bright dipping/pickling operations.	Material substitution Regeneration of bright pickle dip using electrolytic recovery or chemical precipitation.
3. Nickel and chrome-loaded acid baths from immersion stripping operations.	Equipment redesign using electrolytic recovery to regenerate bath.
4. Solvents from degreasing operations.	Material substitution Recovery involving carbon absorption and/or distillation. Equipment redesign.

tained. A contaminated plating solution will increase the hazardous materials in dragout and will cause inefficient and defective plating. Eventually, the contaminated solution, if not cleaned, will have to be managed as a hazardous waste. Preventing this from happening is essential before any of the other waste reduction solutions for dragout can be considered. Contamination of the solution can come from a number of sources: metallics from dropped parts; organics from oils or buffing compounds; contaminants present in the municipal water supply; and bleedout from poorly maintained racks or coatings and other objects.

Keeping the plating solution clean involves standard practices in the industry: electrolytic purification in which metal contaminants are plated to a cathode, known as "dummying"; pH adjustment in order to precipitate out metal contaminants; and carbon filters and peroxide treatment to remove organic contaminants. The solutions used to clean the parts before they are plated can also be recycled using filtration. The filter media used will depend on what materials are contaminating the solution; these usually include buffing compounds, oils, or hexavalent chromium.

However, filtration itself can create disposal problems for the plater. These problems occur when the filters become saturated and must be changed. During filter changes, some of the plating solution may be spilled; not only is the valuable solution lost, but also more hazardous waste is produced.

In order to reduce these sources of waste, Standadyne, Inc., a plumbing products manufacturer in Sanford, North Carolina, instituted a new filtration system in which low-pressure air is blown into the filter through the bleed-off hose (see Exhibit 5). The air blows the plating solution out of the filter and into a slurry or storage tank. In this way, the filter becomes dry and thus much easier and safer to change, and little or no solution is lost. In addition, Stanadyne switched from sleeve-type filters to new, horizontal plate filters, manufactured by Baker Brothers, Inc. in Jacksonville, Florida. The new filters require fewer changes per unit volume of solution filtered, and thus few of them have to be purchased.

Raw Material Conservation. Reducing the metal concentration of the plating bath can not only reduce dragout but also save the company money in raw material costs. How much of the metal in solution a company is able to reduce depends on the physical properties of the parts and the company's plating quality requirements.

Stanadyne also found that it could lower the concentration of the constituent chemicals in its cyanide copper and chrome plating baths without affecting the plating quality. The potassium cyanide concentra-

Exhibit 5. Blowdown of filter.
Source: "Enviroscope," *Plating and Surface Finishing*, June 1985.

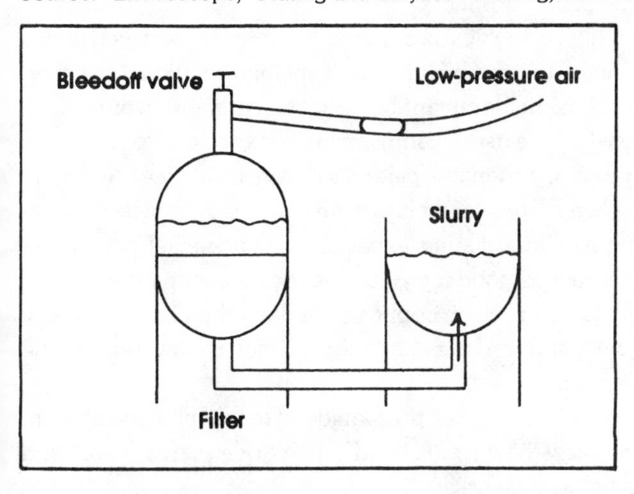

tion in the cyanide-copper bath is run at 2.5 ounces per gallon instead of 3.5 ounces per gallon; this has reduced the dragout concentration by 28 percent. In addition, Stanadyne uses less cyanide for plating and less bleach for cyanide destruction.

In the chrome plating baths, chromic acid levels are maintained at 29 ounces per gallon instead of 32 ounces per gallon. This has saved some of the chromic acid, as well as some of the chemicals used to treat the chromium, such as sodium bisulfite and sodium hydroxide.

Material Substitution. Substituting the ordinary plating solutions with less hazardous varieties can substantially reduce dragout concentration, and in many cases have no detrimental effect on plating quality or efficiency.

Hexavalent chromium can be replaced with trivalent chromium in many operations. Trivalent chromium has a lower viscosity, meaning that less of it clings to the part, and consequently less dragout is created. Trivalent chromium is also less hazardous than hexavalent chromium, posing a lower risk to the environment. However, a company's process may prohibit the use of trivalent chromium, particularly if a certain product quality is required. Trivalent chromium tends to produce a slightly darker surface than hexavalent chrome. However, with proper re-education, most specifications can be satisfied with trivalent chrome-plated products.

Cadmium, which is highly toxic, can sometimes be replaced with a far less toxic zinc-nickel combination in the plating solution. However, the military, which is the dominant buyer in the cadmium-plating field, continues to insist on the use of cadmium in their specifications. Unless these specifications are changed, platers will have no choice but to continue to use the highly toxic cadmium instead of the zinc-nickel substitute. According to some plating experts, the zinc-nickel plating approach produces just as good a product as the cadmium approach.

Copper sulfate can replace copper cyanide in the plating of many metals. Steel, zinc, and lead are exceptions, which usually require the copper cyanide solution.

Acid tin chloride can replace tin cyanide in the tin plating solution. Aside from being less hazardous, acid tin chloride plates better and faster than the tin cyanide original.

Equipment Redesign. The racks which hold the parts during plating can also contribute to dragout. Reducing the surface area of the rack can reduce dragout by providing less area for the plating solution to adhere to. Stanadyne adopted this strategy during its rack redesign, in which it changed from a horizontal to a vertical orientation of its part support members.

Most racks are designed in the classic style of a rectangular box, in which parts to be plated are racked on tips anchored to horizontal crossmembers. In a vertical orientation (see Exhibit 6), the horizontal crossmembers are replaced with vertical supports, allowing the plating solutions to slide off the rack instead of clinging to the extra, horizontal surface area. Another benefit from this vertical design is that there are fewer points of electrical resistance, since the splines are connected directly to the rack hooks and the tips connected directly to the splines. This means that more parts will remain where they belong, on the rack, instead of being accidentally dropped into the plating bath where they can cause contamination. Recoating costs for racks are lower because there is less surface area to recoat.

Aside from the rack itself, the orientation of the part on the rack can also be changed to minimize dragout:

Exhibit 6. Horizontal (left) and vertical rack design.
Source: "Enviroscope," *Plating and Surface Finishing*, July 1985.

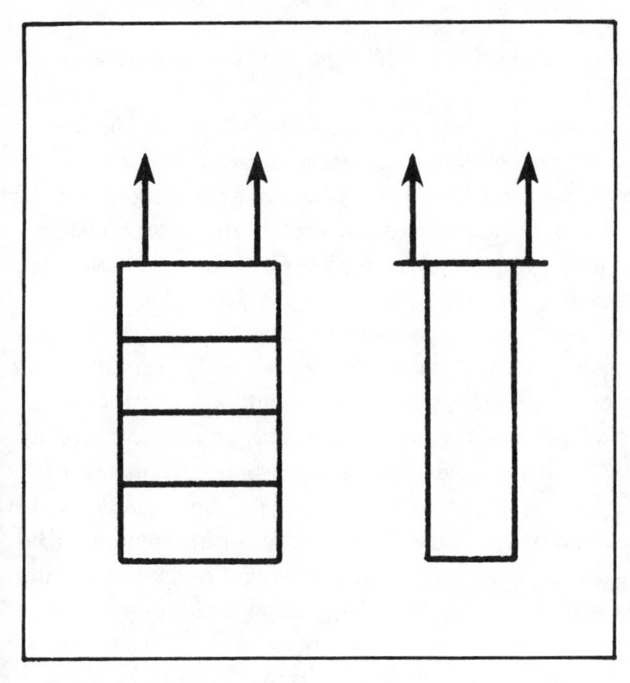

- Parts can be placed so that "cups," "blind holes," and wide shelf areas are avoided as much as possible;
- The rack can be arranged so that parts on top do not drip on those below;
- The parts can be tilted to improve their drainage; and
- The longest lines of the part can be racked in a vertical plane, so that the dragout will slide off more easily.

Another equipment redesign idea is to have the parts hang for a moment after they've been removed from the plating solution. This allows much of the dragout clinging to the part to drip back into the tank. On an automatic, return-type machine such as the Udylite Cyclemaster, a pause is introduced when the part is in the "up" position, before it's indexed over and lowered into the next tank. However, care must be taken to avoid an excessive "hang-time," which could cause dry-ons or passivation of the solution on the part. On manual plating lines, drip bars can be installed on which the rack is hung and allowed to drain over the plating tank.

Simple Recovery of Dragout. Even if all of the above methods to reduce dragout are implemented, *some* volume of solution will inevitably cling to the parts and plating racks. When these parts are rinsed, the rinse water becomes contaminated with metals and other hazardous materials. There are, however, simple, low-cost ways to recover the dragout and recycle it back into the plating solution.

A rinsing system using a high-pressure spray will not only use less water than ordinary rinses, but can also produce a solution with a high concentration of the plating material. This solution can either be returned to the plating bath for reuse or passed through an ion-exchange system to recover the metal, which is then returned to the plating bath. Jim Adams, environmental engineer at National Cash Register (NCR) in Columbus, Ohio, is currently using a spray rinsing system for his printed circuit boards, and he reports a 90 percent reduction in water usage and a 94 percent recovery of nickel metal, which is returned to the plating bath. How well this system can be applied to the electroplating industry depends on the particular process a plater is using; passivation of the nickel may prevent or restrict its use in some cases.

Another low-cost recovery method is to place a dragout recovery tank at the end of the plating line, directly after the last plating tank (see Exhibit 7). The recovery tank collects additional dragout that drips from

Exhibit 7. Dragout recovery tank for nickel plating on steel parts.
Source: "Enviroscope," *Plating and Surface Finishing,* August 1985.

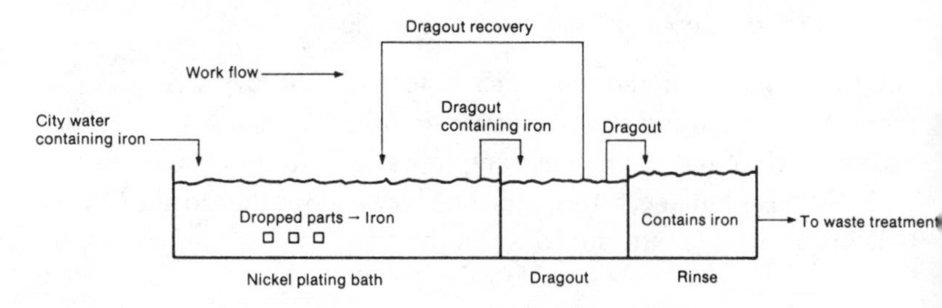

the parts; the contents of this tank are returned directly to the preceding plating tank. However, this system is practical only for plating solutions that are operated at high temperatures; this way, the recovered dragout can be used to replenish the plating solution as it evaporates. Otherwise, the metal concentration in the dragout recovery tank will build up and eventually interfere with the tank's rinsing ability. Systems that use plating solutions with a low evaporative loss rate (such as decorative chromium) must either reduce the solution volume in the dragout tank or increase the evaporation rate in the plating solution. Many companies now sell room-temperature evaporators that use large surface areas to increase the evaporation rate.

Other alternatives in dragout recovery include dewatering the rinse stream and precipitating the metal ions out of solution. Stanadyne purchased a J-Mate sludge dryer manufactured by JWI in Holland, Michigan, which drives nearly all of the water out of its metal sludge. The remaining solids have a metal content of 30 to 40 percent; these metals usually can be recovered.

Pioneer Metal Finishing, Inc. in Franklinville, New Jersey, adds chemicals to its storage tanks in order to precipitate the metals out of the dragout solution. The water decanted off the top can be reused, except for rinses that require high quality, distilled water. Pioneer plans to sell its metal sludge to a concrete manufacturer for use as a filler in concrete products.

2. CHROMIC ACID AND COPPER IONS

Material Substitution. A bright dipping or "pickling" operation cleans the surfaces of brass and copper parts with the aid of chromic acid and oxidizers. The spent chromic acid, a hazardous chemical, is responsible for a large amount of water that platers must dispose of in a hazardous landfill. Replacing the chromic acid with nonhazardous cleaning chemicals can substantially reduce the toxicity of the waste stream.

Elkhart Products, Inc., a pipe fittings manufacturer in Elkhart, Indiana, installed a new bright dipping system which uses a sulfuric acid and hydrogen peroxide pickle dip instead of the conventional chromic acid dip. The sulfuric acid and hydrogen peroxide are much cheaper to obtain than chromic acid; consequently, Elkhart is saving $1,770 per year in raw material costs and an additional $18,200 per year in avoided waste disposal costs. Stanadyne, Inc. of Sanford, North Carolina, made a similar substitution in their bright dipping line and also realized a substantial reduction in their operating costs.

Electrolytic Recovery. As the bright pickle dip cleans the copper metal surfaces, it eventually becomes saturated with copper ions, which inhibit its cleaning ability. Instead of throwing out the dip as that much more waste to be disposed of, it can instead be cleansed of the copper ions, using either electrolytic recovery or chemical precipitation. If electrolytic recovery is used, platers can actually save substantial amounts of money from the process by removing the copper metal from the cathode and selling it as number 1 copper scrap.

Elkhart Products in Elkhart, Indiana, uses both electrolytic recovery and precipitation. Their treatment system, planned and installed by the consulting firm of Lancy International, precipitates the copper out of the pickle dip as copper sulfate; the rinse water is also regenerated, precipitating the copper out as cuprous oxide and copper hydroxide. The reason that Elkhart uses precipitation at this stage is so that the pickle dip and the rinse water can be kept separate for reuse. The precipitates are combined and introduced into an electrolytic cell, which plates out the copper at a rate of 1,000 to 1,500 pounds a month. Stanadyne, Inc., on the other hand, uses electrolytic recovery directly on its dip. In both instances, copper is recovered for reuse instead of being disposed of as a hazardous waste.

3. NICKEL AND CHROME-LOADED ACID

Stripping nickel and chrome off brass parts is a rework operation designed to save the expensive brass base metal. Immersion stripping in an acid bath is the most commonly used method to remove the nickel and chrome. The acid bath loses its stripping efficiency as it becomes loaded with the stripped metal, and usually it must be treated as a hazardous waste, and a new stripping solution must be prepared.

Equipment Redesign. Stanadyne, Inc., in Sanford, North Carolina, solved this problem by purchasing a new, "Udystrip 8000" electrolytic nickel stripper, manufactured by the Udylite division of OMI International in Warren, Michigan. The new equipment constantly regenerates the acid bath by plating the stripped nickel onto electrodes. Nickel is eventually removed from the electrodes and sludge is removed from the bottom of the tank. The contents of the bath are then pumped back into the tank. A volume of liquid, equal to the volume of the sludge removed, is added to the batch to return it to former levels. Stanadyne reports that the new stripper has resulted in an 81 percent reduction in chemical costs per hour, saving approximately $5,000 per year.

4. DEGREASING SOLVENTS

The waste reduction solutions available for degreasing operations are discussed in-depth in Chapter 5, Nonelectric and Electric Machinery Manufacture (see page 70). These solutions include: equipment redesign, involving the installation of chillers, covers, and freeboard extension units; distillation, either on-site or by an off-site recycling firm; and material substitution of a water-soluble cleaner for the solvent cleaner usually employed. In addition, cyanide cleaners, which platers also use in degreasing operations, can be replaced with trisodium phosphate or ammonia, which are especially powerful when used in a hot, ultrasonic bath.

Equipment Redesign. In an approach similar to the one taken toward their immersion stripping operations, Stanadyne determined that their degreasing operations could also be upgraded and improved. After analyzing all the latest machines available on the market, the company decided to go with a fully enclosed, automatic degreaser manufactured by Finishing Equipment, Inc. of St. Paul, Minnesota, because it incorporates a number of pollution-prevention features as part of its process. Some of the machine's most attractive features include the following:

- A chiller system which runs continuously in order to keep the solvent in the machine, where it belongs. Major solvent losses often occur when a degreaser is not in operation. By running the chiller system continuously, these losses are eliminated.

- A cooling and rinsing system which does not need water. This reduces water costs.
- A conservation system which uses waste heat from the chiller to heat the degreaser unit, thus saving energy.
- A continuous rotation of parts in order to facilitate drainage and prevent solvent dragout.
- A degreaser system which uses methylene chloride, a chemical whose stability allows it to be distilled many times before it degrades.
- A large distillation capacity, which can reduce still bottoms to a 6 to 10 percent solvent content.

This recently purchased degreaser is already paying dividends for Stanadyne. Solvent usage has dropped from an average of 7 drums per month to only 1.5 drums per month, and air emissions and hazardous waste generation have been reduced substantially.

Contacts For Further Information

1. George F. McRae, CEF
   Plating/Pollution Control Manager
   Stanadyne, Inc., Moen Group
   Sanford Plant, Cox Mill Road
   Sanford, North Carolina 27330
   Contributing expertise: Dragout reduction and recovery.
2. Dave Gibson
   Baker Brothers, Inc.
   7892 Baymeadows Way
   Jacksonville, Florida 32216
   Contributing expertise: Manufacturer of horizontal plate filters.
3. Jim Adams, Environmental Engineer
   National Cash Register (NCR)
   P.O. Box 728
   Columbus, Ohio 43725-0523
   Contributing expertise: Spray rinse system for dragout recovery.

4. JWI
   2155 112th Avenue
   Holland, Michigan 49423
   Contributing expertise: Manufacturer of J-Mate sludge dryer.
5. Harry DeSoi, President
   Pioneer Metal Finishing, Inc.
   Colesmill Road
   Franklinville, New Jersey 08322
   Contributing expertise: Chemical precipitation to recover dragout.
6. Elkhart Products Corporation
   1255 Oak Street
   Elkhart, Indiana 46515
   Contributing expertise: Electrolytic recovery and chemical precipitation to regenerate bright pickle dip.
7. William McLay, Consultant
   Lancy International
   525 West New Castle Street
   P.O. Box 490
   Zelienople, Pennsylvania 16063
   Contributing expertise: Consultants to Elkhart Products Corporation.
8. OMI International
   Udylite Division
   21441 Hoover Road
   Warren, Michigan 48089
   Contributing expertise: Manufacturer of "Udystrip 8000" electrolytic nickel stripper.
9. Finishing Equipment, Inc.
   3640 Kennebec Drive
   St. Paul, Minnesota 55122
   Contributing expertise: Manufacturer of a fully enclosed, automatic degreaser.

5

Nonelectric and Electric Machinery Manufacture (SICs 35 & 36)

1. CHLORINATED SOLVENT CLEANERS

Equipment Redesign. Used chlorinated solvent cleaners, such as 1, 1, 1-trichloroethane, are by-products of degreasing metal parts prior to painting or assembly. In most cases, a company's first step in reducing the production of these waste solvents is to examine the degreasing equipment to see if any modifications can be made to prevent solvent evaporation. Three such modifications include: (1) raising the freeboard height to 75 percent of the tank width instead of the conventional 1/2 tank width; (2) placing covers on open-tank systems whenever the system is not in use; and (3) installing condensing coils, known as "chillers," near the top of vapor degreasers (see Exhibit 8).

Pierce Industries, Inc., a metal furniture manufacturer in Walden, New York, installed a "Detrex" freeboard extension unit and an automatically closing cover on their degreaser. In the first six months of operation, these modifications cut solvent use by 1,650 gallons, saving

Waste-Producing Product or Process	*Waste Reduction Solution(s)*
1. Chlorinated solvent cleaners used in degreasing operations.	Equipment redesign Material substitution. Recycling involving distillation.
2. Acid, caustics, metal, and wastewater from electroplating operations.	Equipment redesign (and other solutions from related section in Chapter 4).
3. Paint solvent sludge and vapors from painting and drying operations.	Equipment redesign or material substitution. Recovery using carbon absorption or nitrogen-based drying.
4. Oil and other process chemicals from machining operations.	Recovery involving ultrafiltration. Chemical treatment and recovery.

the company $6,600 in virgin solvent purchases. Sealed Power Corporation, a piston ring manufacturer in Muskegon, Michigan, installed chillers on seven of their degreasers. The company estimates that two of the chillers alone saved them $17,688 a year in solvent costs; this prompted them to install chillers on their remaining degreasers. (For more information, see Campbell and Glenn, 1982.)

Material Substitution. In some applications, the organic solvent degreaser can be replaced by a water-soluble synthetic cleaner, depending on the corrodibility of the metal parts. The water-soluble cleaner is usually less expensive than the organic one and presents no disposal problems because it is nonhazardous.

The Hamilton Beach Division of Scovill, Inc., a small appliance manufacturer in Clinton, North Carolina, purchases a water-soluble cleaner from the Cincinnati Milacron Company of Cincinnati, Ohio.

Exhibit 8. By retrofitting an open-top vapor degreaser with second set of condensing coils, a cold blanket of air is created above the vapor zone, preventing the escape of solvent vapors.

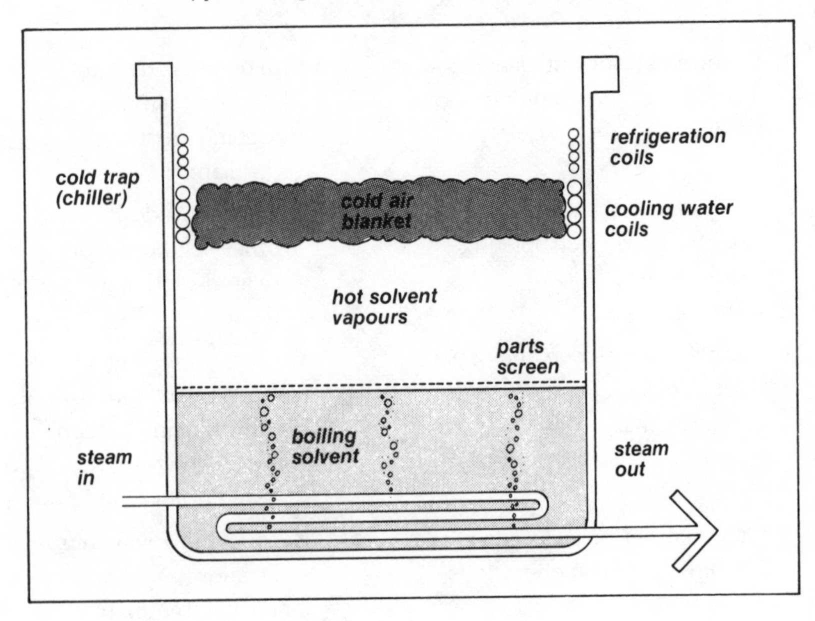

Scovill is able to reduce its 1, 1, 1-trichloroethane use by 30 percent and reports a $12,000 per year savings from the substitution.

Distillation Recovery. Another alternative is to recycle the spent solvent using *distillation recovery techniques.* Because of their low boiling points, most degreasing solvents can be distilled very easily. A typical distillation unit recovers 85 to 90 percent of the original solvent.

A company must face the decision of whether to invest in its own distillation unit or to contract the services of an off-site recycler. There are advantages and disadvantages to both choices.

An off-site recovery arrangement eliminates the capital expenditures needed for distillation equipment, as well as the maintenance and operating costs associated with such equipment. On the other hand, the generator must assume the liability for his waste during transport to

Photo 3. Detrex solvent degreasing tank with powered tank cover.

Detrex Chemical Industries Inc.

the off-site facility. Off-site recycling can also sometimes result in contamination of the recovered solvent with a substance from another user's process.

Scovill contracted the Ashland Chemical Company to recycle 27 drums of accumulated solvent waste as well as all future waste, estimated to be about 38,000 pounds per year. Substituting the recycled solvent for the virgin product reduced raw material costs from $0.41 per pound to $0.27 per pound, yielding an annual savings of $5,320 per year. Recycling also enabled Scovill to eliminate all of their previous waste disposal costs, estimated to be about $0.08 per pound, or $3,040 per year. Within six months of deciding to recycle, Scovill was able to realize a payback of $8,360.

On the other side of the coin, Rexham Corporation, a printing

Photo 4A. Small-scale batch distillation unit suitable for recovering common solvents such as acetone, alcohols, chlorinated solvents, and Freons.™

operation in Greensboro, North Carolina, purchased a Cardinal distillation unit for in-house recovery of alcohol/acetate ink solvent waste. The total cost of the unit and the required structural modifications were $16,000. The unit recovers 85 percent of the solvent from the waste stream, resulting in $15,000 saved per year in virgin solvent costs and $22,800 saved in hazardous waste disposal costs.

Rexham must still send 72 drums of solvent residue a year to a hazardous waste landfill, although this is far less than their previous 300 to 350 drums per year. In addition, still bottoms are transported for

Photo 4B. The Finish Little Still LS-15 can recover 75 to 100 gallons of solvent per week.

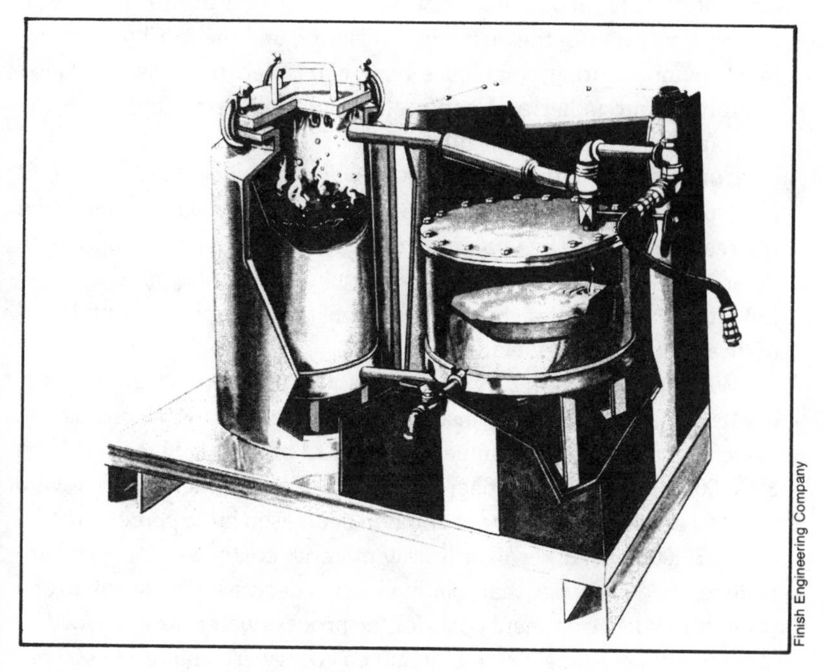

off-site incineration at a cost of $98.50 per drum. Rexham plans to use the money saved from the Cardinal unit to buy a "second generation" solvent reclamation unit distributed by Activation, Inc. of Charlotte, North Carolina. The new unit has a more efficient heating element, which allows almost total recovery of solvent. The residue from the second unit will be nonhazardous and thus will be sent to a sanitary landfill, allowing the Greensboro plant to be declassified as a hazardous waste generator.

2. ELECTROPLATING WASTES

Electroplating operations have a tradition for being big producers of highly toxic hazardous substances, including acids, caustics, cyanide, and metals such as zinc, copper, and chrome. In addition, a large amount of contaminated wastewater is produced from rinsing procedures. Because of the high cost of disposal and the liability involved, many manufacturing companies have eliminated their electroplating operations altogether and contracted electroplating firms to do the work (and assume the liability) for them.

Equipment Redesign. For those who have decided to keep their electroplating operations in-house, there are many ways to minimize the resulting waste [See Fabricated Metals (Electroplating) for a more detailed discussion]. Equipment redesign can not only reduce waste (and treatment costs), but also streamline the operation and dramatically increase productivity.

When the Emerson Electric Company, a power tool manufacturer in Murphy, North Carolina, installed an automated metal electroplating system to replace their manual one, annual productivity increased by $200,000 and down-time decreased 50 percent. Other benefits were realized as well: chemical consumption decreased by 25 percent, resulting in $8,000 saved per year in raw material costs; water costs were reduced by $1,100 per year; plating wastes decreased from 450 to 360 pounds per day; treatment costs for the process water were reduced by 25 percent; personnel and maintenance costs were cut by $35,000 per year; and worker exposure to acids and caustics was eliminated, an important factor in the company's medical insurance bills.

Emerson made an initial investment of $158,000 in the new system, which paid for itself (and even turned a profit) in little more than a year. However, the price of such systems is going up: According to Layton Schuh, chemical engineer for Emerson, the same electroplating equipment the company purchased in 1978 for $158,000 costs $350,000 in 1986.

3. PAINT SOLVENTS

Most metal products require one or more coats of paint, not just for appearances, but also to protect metal parts from corrosion. The waste streams emanating from the painting process include: (1) excess contaminated paint, containing hazardous solvents as its base (a liquid waste stream from the application process), and (2) solvent vapors, a product of curing the painted objects in a hot oven. Each of these waste streams has its own solution.

Material Substitution/Equipment Redesign. Like the chlorinated degreasing solvents, organic-solvent-based paints can in some cases be replaced with water-based paints. Admittedly, some industries require certain kinds of paint to maintain their product quality, and this may preclude their changing to current formulations of water-based paints. (For example, according to Paul Hoffman of Garden Way, Inc., car manufacturers require lead paints for the body because these paints keep their color better under stress.) Generally, however, more and more industries are finding that their coating and painting operations can be performed with water-based formulations.

The Emerson Electric Company has been able to substitute an anodic electrostatic immersion paint system that is water-based for the previous organic-solvent-based system. The result: Emerson actually found that its product quality improved. The new paint system coats the parts more uniformly and more completely. Down-time decreased by 67 percent.

The cost savings are spectacular. Annual productivity increased by $1 million; raw material costs for virgin paint decreased by $600,000 per year because the water-based system allows 99.5 percent recovery and reuse of paint. Instead of generating 3,000 pounds of chlorinated aromatic solvent waste and 70 pounds of waste paint solids per day, the company now produces only 150 pounds of nonreactive solvent and 2 pounds of waste paint solids per day. As a result, hazardous waste disposal costs have decreased from $10,000 per year to only $300 per year. Personnel and maintenance costs have been decreased by 40 percent because of the new system, and worker exposure to harmful organic paint solvents has been eliminated.

Solvent Vapor Recovery. Two technologies, one old and one new, are available to cut solvent losses during the drying phase. One of these is carbon absorption, which is well known in industry and has been around for at least 40 years. It consists typically of three stages:

1. A bed of activated carbon absorbs solvent vapors from the passing air.
2. The saturated carbon bed is stripped of the solvent temporarily attached to it; this is known as "desorption." Desorption can be

accomplished by several methods: low pressure steam; a solvent which removes the absorbate; acidic or caustic wash followed by thermal activation of the carbon; and indirect heating with a nitrogen purge. Steam is most frequently used because it is comparatively inexpensive, easily condensed back to a liquid, and doesn't present a disposal problem in itself.

3. The solvent in solution is recovered, either through simple decantation if the solvent is immiscible in water, or otherwise through distillation.

Some spokesmen for the industrial painting and coating industry question whether carbon absorption is economically feasible, especially for smaller operations. They cite two problems: (1) often there is a mixture of solvents in the paint, which requires an expensive distillation technology to recover the individual components; and (2) the solvents are typically in a low concentration in a large quantity of air, meaning that very little solvent is recovered per unit volume of vapor.

A new technology in drying operations is being marketed by Airco Industrial Gases in Murray Hill, New Jersey. Whereas most conventional drying ovens need a constant source of air in order to dilute the solvent vapor below explosive levels, Airco's new system replaces air with an inert nitrogen atmosphere. The nitrogen allows the solvent vapor to exist at much higher concentrations, and the system is entirely self-contained. Airco claims that the nitrogen-based system is more energy and recovery efficient.

The equipment was tested at a fabric-coating facility in Kenyon, Rhode Island, and was found to reduce heating energy requirements by 60 percent over the previous system, saving $7.75 an hour. And because the solvent is at a much higher concentration, it can be recovered directly by condensation; this translates into a 99 percent recovery efficiency. However, this new technology may not be practical at the present time for small and medium-sized businesses because of its high capital cost (For more information, see Campbell and Glenn, 1982).

4. OIL AND OTHER CHEMICALS

Ultrafiltration Recovery. Used oil and other chemical byproducts from the manufacturing process can be recovered from the waste stream using *ultrafiltration* techniques. Ultrafiltration involves passing the waste stream through a membrane filter that retains the oil molecules while allowing other, lower weight molecules to pass through. The membrane can be made of various substances, including cellulose nitrate, cellulose acetate, polysulfone, aramids, polyvinylidene fluoride, and acrylonitrile polymers. Oil skimmers, on the other hand, recover oil from holding tanks in the same way that cream is skimmed from milk.

The Emerson Electric Company installed a $65,000 oil skimmer and ultrafiltration unit, which recovers $8,000 worth of oil per year. The unit also recovers 25,000 gallons of reusable alkaline cleaning solution each month, saving approximately $3,000 each year. Because the unit reduces organic loading to the water treatment system, treatment costs have been cut by $10,000 per year, and the company has been spared the expense of expanding their water treatment capacity.

Chemical Treatment and Recovery. John Deere and Company, a farm and construction equipment manufacturer in Moline, Illinois, has a liquid waste treatment facility which reclaims oil as part of its processes. Emulsion breakers, anionic polymers, and acids are added to the oil, which is then "cooked" at 180 degrees Fahrenheit for 12 hours, producing a good quality reclaimed oil. Over a million gallons each year of waste oils and oil/water mixtures are treated in this way; the recovered oil is reused in the machining processes or sold to an oil recycling firm. Deere and Company has a patent pending on the process, but is willing to license its use to other industries. For more information, contact Mike McGuire at John Deere and Company, cited below.

Contacts For Further Information

1. Sealed Power Corporation
   100 Terrace Plaza
   Muskegon, Michigan 49443
   Contributing expertise: Equipment redesign of solvent degreasers.

2. Guy Tilford
   Hamilton Beach Division of Scovill, Inc.
   P.O. Box 70A

Clinton, North Carolina 28328
Contributing expertise: Material substitution and distillation recovery of degreasing solvents.

3. Cincinnati Milacron, Inc.
   4701 Marburg Avenue
   Cincinnati, Ohio 45209
   Contributing expertise: Manufacture of water-soluble degreasers.

4. Ashland Chemical Company
   3930 Glenwook Drive
   Charlotte, North Carolina 28208
   Contributing expertise: Solvent recycler contracted by Scovill, Inc.

5. Gordan Miller, Manager of Safety and Environment
   Rexham Corporation
   P.O. Box 368
   Matthews, North Carolina 28106
   Contributing expertise: In-house distillation recovery.

6. Activation, Inc.
   8041 F Arrowridge Boulevard
   Charlotte, North Carolina 28210
   Contributing expertise: Manufacturer of solvent reclamation units.

6. Layton Schuh, Chemical Engineer
   Special Products Division
   Emerson Electric Company
   2001 Highway 64 East
   Murphy, North Carolina 28906
   Contributing expertise: Equipment redesign of electroplating and painting systems. Installation of oil skimmer and ultrafiltration unit to recover oil.

7. Mike McGuire
   John Deere and Company
   John Deere Road
   Moline Illinois 61265
   Contributing expertise: Chemical treatment and recovery of oil waste.

6

Electronic (Printed Circuit) Equipment (SIC 36)

1. ETCHING AND PLATING BATHS

While competition is keen among most industrial sectors, it is especially keen among the members of the printed circuit board and related micro-electronics industries. Printed circuit manufacturers are exploring and implementing waste reduction solutions whenever they make good economic sense and at the same time do not adversely affect product quality or productivity. Fortunately, the waste reduction solutions described below can recover valuable and expensive raw materials, such as copper, and in some cases, even enhance productivity and product quality.

Electrolytic Recovery. Officials at GTE Sylvania's plant in Chicago, Illinois, which manufactures telephone switching equipment, recently decided to change their single-pass waste treatment system into a closed-loop recovery system. The company employed the consulting firm of Lancy International to help plan an overall waste reduction program. Part of this involved the installation of an electrolytic metal recovery cell in the company's "Electroless" copper plating line.

The negative cathode of the electrolytic cell attracts the positive

Waste Producing Product or Process	*Waste Reduction Solution(s)*
1. Etching and plating baths contaminated with heavy metals, especially copper.	Electrolytic recovery Closed-loop batch recovery involving chemical treatment.
2. Acids from etching and plating operations. Still bottom waste from solvent reclamation.	Atmospheric evaporation recovery. Waste exchange.
3. Chemicals used in the cleaning of the copper metal.	Equipment redesign.
4. Rinse water contaminated with metal ions from rinsing operations.	Equipment redesign using a spray rinse system.

copper ions in the copper etching rinse stream as it flows out of the rinsing tank. The copperless rinse is then recirculated to the tank for reuse in the rinsing operations. Periodically the copper metal is removed from the cell's cathode and sold to copper processors.

At first, GTE's recovery efforts were motivated by the stricter regulations on copper effluent soon to be enforced by the Chicago Metropolitan Sanitary District. Now the company is finding that their recovered copper can be sold as number 1 copper scrap, generating $2,000 in revenue each year. In addition, GTE avoids $4,000 in landfilling costs.

A division of Data General Corporation in Clayton, North Carolina, which manufactures printed circuit boards, is pilot-testing some new electrolytic equipment manufactured by Metal Removal Systems in Long Island, New York. Stan Taylor, chemical engineer for Data General, says that a major advantage of the new system is that it can plate the copper down to discharge limits, achieving almost total removal of the copper from solution. This way, the company doesn't have to blend any electrolytic by-product back into the waste stream in order to comply with effluent guidelines. The system accomplishes this through two features: (1) a captive electrolyte, which doesn't generate any liquid waste solution during cathode stripping; and (2) a woven fiber cathode, which provides more surface area for plating than provided by most cathodes.

In contrast to GTE Sylvania, Stan Taylor uses a different setup at Data General for recovering copper from the rinse stream. Instead of using the electrolytic cell to continuously remove copper ions from the closed-loop flow, Taylor saves the electrolytic process for the end of the day, when the rinse stream is highly concentrated with the copper ions. "The advantage of this system," Taylor says, "is that it prevents the build-up of unwanted material on the cathode." At the end of the day, ions other than the copper can be precipitated out of solution using sodium hydrosulfate, leaving only copper ions to be plated.

In actuality, electrolytic recovery is more widely used on the metal-loaded etching and plating baths from printed circuit manufacture rather than on the rinse streams. This is because etching and plating baths have a much higher concentration of the copper ions, making electrolytic recovery that much more effective and productive. With rinse streams, electrolytic recovery is mainly used to make them suitable for release into the environment.

Chemical Precipitation. Some printed circuit manufacturers continue to rely on chemical precipitation to recover their copper from solution rather than on the more advanced method of electrolytic recovery. While methods vary widely, chemical precipitation usually produces an impure copper sludge, whereas electrolytic recovery produces copper that is usually 99 percent pure.

Data General precipitates copper out of solution because its chemical treatment process achieves a 90 percent concentration of metallic copper. Spent copper-bearing acid baths, such as sulfuric-peroxide etchant, are added to the wastewater in order to raise its pH, and formaldehyde is also added as a reducing agent. The copper in solution is thus destabilized. When finely-powdered copper sludge, generated in a previous process, is added, the destabilized copper precipitates out as elemental copper, which is subsequently sold to markets which can use the impure form of the metal. As a result, 151,800 pounds of hazardous sludge are eliminated each year.

A major advantage of this process is that different waste streams, including the etchant and powdered copper sludge, can be combined to detoxify each other and produce a reusable copper, rather than having to bury this previously wasted resource. Data General's modifications paid for themselves in 1 1/2 months, saving the company $180,000 per year in avoided disposal costs and recovered raw material.

2. ETCHING AND PLATING ACIDS

Atmospheric Evaporation Recovery. Evaporation is a simple recovery technology which can be easily applied when the atmospheric conditions are right. A dry, hot climate is most conducive to evaporation. For example, in Tempe, Arizona, Digital Equipment Corporation, a printed wiring board manufacturer, uses an atmospheric evaporator to recover the chromic acid used in its etching operations. The evaporator recovers approximately 95 percent of the acid, which is then reused in the etch line. Digital Corporation, like GTE, also used Lancy International to help them plan their waste reduction program.

Waste Exchange. Another alternative is waste exchange, provided that a company can find markets for its waste. Data General Corporation separates its etching and plating waste streams into those from which copper can be recovered through electrolytic methods, and those which can be sold directly to companies that use the chemicals. Its salable etchants are cupric chloride and ammonia, and the salable plating acids include acid copper, palladium catalyst, sulfuric acid, and spent nickel-plating baths. Originally, the cupric chloride and acid copper were sold because Data General didn't have its electrolytic equipment in place, and the buyer had the facilities to recover the copper. This might be an inexpensive alternative for companies that cannot afford to put electrolytic equipment in place.

Some byproducts of recovery processes are also salable, such as copper sulfate crystals from the regeneration of the sulfuric-peroxide bath, and still bottoms from 1, 1, 1-trichloroethane recovery (see section 1 from Nonelectric and Electric Machinery). Data General ships 90,000 pounds of wastes each year to buyers, either at no cost or for profit, which has yielded the firm a total of $9,000 per year in combined disposal cost savings and revenue.

3. CLEANING AGENTS

Equipment Redesign. Some electronic manufacturers require a clean copper sheet metal before it can be used to make the electronic circuits. Normally the metal is cleaned with acid-based chemicals. 3M's microelectronics plant in Columbia, Missouri, which manufactures flexible electronic circuits, previously cleaned its copper metal by spraying it with ammonia persulfate, phosphoric acid, and sulfuric acid. The process produced hazardous waste that was expensive to manage properly. The new process is entirely mechanical: A machine with rotating brushes cleans the metal with fine abrasive pumice, leaving a nonhazardous sludge that can be placed in a conventional sanitary landfill. As a result, the plant's total production of hazardous waste is reduced by 40,000 pounds a year. In addition to easing the disposal problem, the new method lowers raw material and labor costs, generating a net savings of $15,000 each year. The new equipment cost $59,000 to acquire, but the savings it generates should pay back the investment in three years.

4. CONTAMINATED RINSE WATER

Normally the finished printed circuit board is rinsed by immersing it in a water bath; eventually this bath becomes contaminated with metal ions and other contaminants from the board. This method uses a lot of water and produces a lot of waste, as each contaminated bath must be managed as a hazardous waste and a new solution prepared.

Equipment Redesign. Jim Adams, environmental engineer at National Cash Register (NCR) in Columbus, Ohio, came up with a simple but effective system to clean the boards using much less water. He subjects his products to four high-pressure water sprays, which achieve 99.7 percent removal of contaminants from the boards. The advantage of a spraying system instead of a soaking system is that it can reach "cups" and "holes" in the products using a relatively low volume of water. Most printed circuit manufacturers require 7 to 10 gallons of water per minute in their rinsing operations. Jim Adams' method uses only a quarter of a gallon per minute, 10 percent of what most manufacturers are using. The rinse stream resulting from the spray contains a high concentration of metal ions; this makes it feasible to use ion-exchange to remove the metal and, after that, reverse osmosis to remove the sulfates. The metal is then available for reuse in the plating line.

Contacts For Further Information

1. Dean Chappell, Safety Engineer
   GTE Sylvania, Inc.
   One Stamford Forum
   Stamford, Connecticut 06904
   Contributing expertise: Electrolytic recovery of copper from etching and plating baths and rinse streams.

2. William McLay, Consultant
   Lancy International
   525 West New Castle Street
   P.O. Box 490
   Zelienople, Pennsylvania 16063
   Contributing expertise: Consultants to GTE Sylvania.

3. Stan Taylor, Senior Process Engineer
   Data General Corporation
   P.O. Box 186
   Clayton, North Carolina 27520
   Contributing expertise: Electrolytic recovery of copper from etching and plating baths and rinse streams.

4. Jim Muir, Manager
   Metal Removal Systems
   70A Carolyn Boulevard
   Farmingdale, New York 11735
   Contributing expertise: Manufacturer of electrolytic equipment.

5. Digital Equipment Corporation
   190 West 14th Street
   Tempe, Arizona 8521
   Contributing expertise: Atmospheric evaporation recovery of chromic acid.

6. Mike Koeningsberger, Environmental Engineer
   Minnesota Mining and Manufacturing Company (3M)
   3M Center
   St. Paul, Minnesota 55101
   Contributing expertise: Equipment redesign of the cleaning of copper sheet metal.

7. Jim Adams, Environmental Engineer
   National Cash Register (NCR)
   P.O. Box 728
   Columbus, Ohio 43725-0523
   Contributing expertise: Spray rinse system for cleaning printed circuit boards.

7

Electric, Gas, and Water Utilities (SIC 49)

1. COAL ASH

Fly ash and bottom ash are by-products of the coal burning process in coal-fired boilers. Although the ash is usually not hazardous, it must still be managed properly. Usually, the ash is sluiced to ash settling ponds, which act as temporary storage sites. As the ponds fill up, new ones must be continually built, adding to the utility's expense.

Waste Exchange. An alternative management strategy is for the utility to donate or sell the ash to markets which can use it: The utility makes a profit and at the same time avoids disposal costs and its associated risks. Because of its pozzolanic property, light unit-weight, and compressive strength, fly ash makes an excellent structural backfill and roadbase stabilizer. Thus, concrete and asphalt manufacturers and transportation departments are all potential markets. Bottom ash can be used as an aggregate in concrete production or in place of crushed stone for secondary roads.

ELECTRIC UTILITIES

Waste-producing Product or Process	*Waste Reduction Solution(s)*
1. Fly and bottom ash from coal-burning generators. Cenospheres from sluicing ponds.	Waste Exchange Incorporation into cement for road and building construction.
2. Polychlorinated biphenyls (PCBs) present in mineral oils of transformers.	Incineration
3. Radioactively contaminated compactible materials, tools, and equipment from nuclear power stations.	Segregation of waste streams. Equipment redesign
4. Radioactive waste products from an evaporator-based treatment of non-recyclable, radioactive waste. Spent boronic acid and water coolant, that is radioactively contaminated but is recyclable.	Equipment redesign using ion-exchange and improved evaporator design.

The success of fly ash and bottom ash sale depends on two important factors: the composition of the ash and the local market options. Ash composition in its turn is affected by a number of factors, including the type of coal burned, the boiler design and its operating conditions, and the methods used to handle and collect the ash. The Electric Power and Research Institute in North Carolina has prepared a by-product utilization manual which contains a step-by-step method for power companies to estimate the feasibility of marketing their ash.

MUNICIPAL SOLID WASTE LANDFILL

Waste-producing Product or Process	*Waste Reduction Solution(s)*
Methane gas from anaerobic decomposition of organic waste in landfill.	Recovery system for collecting the gas, which can then be converted to electric power or distributed for domestic and industrial use.

MUNICIPAL WASTEWATER TREATMENT PLANT

Waste-producing Product or Process	*Waste Reduction Solutions(s)*
Raw sewage sludge, often containing pathogenic organisms and hazardous organic and inorganic chemicals.	Chemical coordinate bond and adsorption system (CCBA) Composting.

Fly ash, which constitutes at least 80 percent of the total ash produced, is collected by electrostatic precipitators as it comes out of the smoke stack. Precipitators can recover 99+ percent of the ash. A vacuum pull removes the ash either to a transfer station or directly to storage silos. Bottom ash is sluiced from the bottom of the steam boiler to holding ponds, where it must be dewatered before it can be used for construction purposes.

Two utilities in North Carolina have been very active in marketing their fly and bottom ash. The Duke Power Company in Charlotte, North Carolina, has been selling its fly ash as a filler for concrete or asphalt since the late 1960's, generating an average of $2.5 million per year in combined revenue and saved waste disposal costs. In 1983 alone, 84,100 tons of fly ash were sold to concrete product manufacturers. Also in that year, 30,000 tons of bottom ash were donated to the North Carolina Department of Transportation for application on secondary roads.

The Carolina Power and Light Company in New Hill, North Carolina, has been selling fly and bottom ash since 1972. Currently, they are not selling the ash directly to users, but instead to two firms: Monier Resources, Inc. of San Antonio, Texas, and Ash Management Corporation of Marietta, Georgia. The two firms are seeking to expand the market for the ash resource. Anywhere from 5 to 35 percent of the total fly ash produced in the U.S. is currently sold, depending on the plant where it's produced. Carolina Power and Light estimates that the initial investment for the ash handling and storage equipment was $1 and $2 per ton, and equipment operation and maintenance expenses are estimated be $3 to $4 per ton.

In some cases, fly ash that has already been disposed of to a pond (cenospheres) can be removed and sold for $400 to $500 a ton to plastic manufacturers for use as a filler. Usually, however, the production cost for drying the ash to make it suitable for sale, together with the transportation costs, don't justify the removal. Cenospheres are very light, and thus the volume in a truck won't produce many ton-units for sale. However, Grover Dobbins at Carolina Power and Light claims that some contractors, such as PA Industries in Chattanooga, Tennessee, will do the reclamation work for cenospheres, as well as provide for its transportation.

2. PCBs

In the 1950's and 1960's, polychlorinated biphenyls (PCBs) were widely used in transformers as a fire-proofing medium. In fact, their use was required at that time by fire departments. Now that the Environmental Protection Agency (EPA) has banned the chemical, PCBs are gradually being phased out as the transformers are cleaned or replaced. However, PCB still contaminates the mineral oil used in transformers, which means that utilities must constantly test the spent oil to see if it exceeds the PCB limit of 50 parts per million(ppm) set by the EPA. If so, the oil must be properly destroyed. One group of transformers, known collectively in the industry as "Akarels," contain transformer fluids with consistently high concentrations of PCBs.

Incineration. The Duke Power company in Charlotte, North Carolina, decided that instead of sending the contaminated oil to a hazardous waste landfill, it would be better to use it as a component of fuel oil for the coal-fired boilers. David Roche at Duke Power says that the oil has a BTU value equal to virgin fuel oil. The contaminated oil is put into the boiler only at high load, when the temperature of the flame is in excess of 2,400 degress Fahrenheit and the temperature throughout the whole system is at least 2,000 degrees, with a residence time of several seconds. In this way, the PCBs are totally destroyed. A temperature of 1,832 degrees is required by the EPA in order for a boiler to operate as a hazardous waste incinerator capable of destroying PCBs.

Duke Power's Riverbend Steam Station, where the PCB destruction takes place, has been approved by the EPA as an incinerator, after stack emission tests showed that the boiler totally destroys the PCBs. The contaminated oil has saved Duke Power $400,000 a year in new fuel oil costs, and $168,000 has been saved in avoided waste disposal costs since 1981. A storage and feed system for the contaminated oil required a $75,000 initial investment. However, the system paid for itself in a few months.

3. CONTAMINATED MATERIALS

Spent radioactive fuel and all materials that have come in contact with such fuel must be disposed of in special facilities designed to contain radioactivity, in sites separate from disposal sites for other hazardous chemicals. Because of the high cost of such disposal, it is essential that power stations using nuclear fission reduce their radioactive waste volume as much as possible. Sometimes this reduction can be simply and easily accomplished.

Segregation of Waste Streams. Compactible materials, such as paper, plastic, and other materials used for radiation protection, make up a large portion of the radioactive waste—as much as 30,000 cubic feet a year. Duke Power's McGuire Nuclear Station in Charlotte, North Carolina, decided to redefine the traditional view of the "radiation area" of the plant.

Contaminated compactible materials are now segregated by area into "absolutely contaminated" and "potentially noncontaminated." Equipment was set up to survey all materials coming from the noncontaminated collection points in order to give final assurance of radioactive cleanliness. "Clean" materials can be used again, and thus the quantity of contaminated materials sent to the landfill is substantially less than the volumes generated in similar plants that do not use segregation. The company purchased and installed improved compactor equipment, which further reduced the volume of waste requiring disposal.

Equipment Redesign. Maintenance tools used on radioactively contaminated components and the components themselves must be disposed of once they are used, producing about 19,000 cubic feet of radioactive waste a year, depending on the size of the plant. The McGuire station determined that a more effective decontamination technology could dramatically extend the lifetime of their tools and lower the amount requiring disposal.

To make sure they would get the best decontamination equipment available, McGuire contracted a number of local service companies, each using a different decontamination technique. The company built up a data base on the effectiveness of each technique.

The new system, manufactured by Allied Nuclear, Inc. of Fremont, California, uses a half-pressure freon spray and freon-based ultrasonics in place of the original chemical soaking vats and water-based ultrasonic system. Previously, the cleansing chemicals were creating disposal problems because many of them were incompatible with the liquid waste process system.

In the new system, a penetrating freon-ultrasonic soak is used as a preliminary step, followed by a freon spray which removes oil, grease, and particulates from the metal surfaces. Previously, electrical equipment could not be cleansed because the old system used water. Investment in the new equipment was $200,000; however, the modifications save Duke Power approximately $1 million per year.

4. CONTAMINATED CLEANING AND COOLING FLUIDS

Equipment Redesign. Converting from an evaporator-based to an ion-exchange-based treatment system for the management of nonrecyclable, radioactive waste, created by equipment drainage and cleaning, is now a general trend in the nuclear power industry. When the McGuire nuclear station installed their ion-exchange system, they were one of the leaders of this trend.

The problem with evaporators is that they're limited in the amount of liquid waste they can handle at any one time. The problem is caused by the nonradioactive components—the borate crystals which precipitate out, build up, and eventually must be disposed of. Evaporators produce large volumes of this concentrate (11,000 to 20,000 cubic feet a year) as a radioactively contaminated by-product. At McGuire, when the evaporators filled up, new waste streams had to be processed by other systems, resulting in increased release of boron and tritium to the environment. The evaporators also lacked the ability to adapt to changes in waste stream chemistry and radionuclide concentrations.

This is precisely the advantage of the ion-exchange system. Ion-exchange resins can be selected to match the chemical make-up of the waste being treated. In this way, only the radioactive ions are removed —the boron salts can be ignored. Thus, there is no salt precipitation build-up and the radioactive by-product, consisting of radioactive ions, is much lower.

Instead of abandoning the evaporator equipment, however, as many utilities are doing as they convert to ion-exchange systems, employees at McGuire found a new use for this equipment. Their evaporators now process the liquid recyclable waste, in particular the boric acid and water used as a coolant to control the fission process.

Before this could be done, the evaporator equipment had to be modified and improved. Several design defects were corrected and 20 new diagnostic instruments were installed, improving the evaporator's performance from 60 to 110 percent of the designed process rate. The evaporator recycles the coolant by boiling off the water, leaving a concentrated boric acid behind. The steam is then condensed as demineralized water. Both recycled products are put through the ion-exchange system in order to prevent the build-up of radioactivity as they are used over and over again. McGuire reports that the ion-exchange and evaporator modifications combined cost approximately $1.3 million, but the savings in avoided disposal costs total $2 million each year.

METHANE FROM MUNICIPAL LANDFILL

A "sanitary" landfill, the type that buries municipal waste, can generate large quantities of methane gas, a product of the anaerobic decomposition of buried waste by bacteria. Many firms are now recovering this methane from landfills and using it in a number of ways. In some cases, it is sold directly to an industrial user, or it is refined to produce a pipeline-grade, high BTU methane for sale to a natural gas company. The Natural Power Company in Raleigh, North Carolina, uses the gas from its sanitary landfill to generate electricity, which is then sold to the local electric utility.

Natural Power "mines" its gas through 10 wells sunk intermittently throughout the Gresham Lake landfill. The wells are constructed of 2-inch thick, schedule-40 plastic pipe, which were sunk "naked" into drill holes running to the bottom of the landfill. Each well is buffered from the landfill contents by a sand and gravel fill, and capped with a 5-to-1 expanding clay and concrete seal. The 15 feet of landfill covering the pipes serve as a buffer to contain the methane in the landfill and to prevent the invasion of oxygen into the anaerobic environment.

The methane gas enters each pipe through 15, 25/1000-inch perforation slots which run the length of the pipe. A 3-horsepower vacuum pump is used to collect the gas through a 4-inch pipe which is connected to each well by a valve. Certain wells produce a consistently higher quantity of methane than other wells, and thus the valve is used to regulate the amount of gas coming from each well, depending on its production capacity. The methane content of the raw gas can vary from 32 to 62 percent; the average is about 42-percent methane.

Two model 3306 Caterpillar diesel engines were modified so that they could use the raw gas as a fuel source. The engines run for 15 hours a day, 5 days a week, and consume 2,684 cubic feet of gas per hour. Because the quality of the gas varies, the engines must be constantly monitored to optimize their performance. Electricity generated by the engines is metered and fed directly into the Carolina Power and Light (CP&L) Company's transmission lines. The engines are operated at their maximum capacity during the peak demand hours when a higher rate per kilowatt hours (kWh) is received. Rates are generally either "fixed," with a higher price per kWh, or "variable." CP&L offers a 15-year contract for fixed rates.

Selling this electricity has generated $25,700 per year for Natural Power. The cost to install the collection system, electrical system, and generators was $125,000, but the methane recovery system should pay for itself in less than five years. However, Bill Roland, president of Natural Power, claims that he has learned a lot from installing the Gresham Lake system; the investment for another operator could be reduced to $100,000.

MUNICIPAL WASTE TREATMENT PLANT

The Chemical Coordinate Bond and Adsorption System. A new, innovative method for treating raw sewage waste, called the chemical coordinate bond and adsorption system (CCBA), is being tested by the San Diego Regional Water Reclamation Agency as an alternative to the traditional decomposition and stabilization techniques. The new system is chemically based as opposed to biologically based, and was developed by the CCBA Resource Center in St. Paul, Minnesota, in cooperation with George C. Harrison, Ph.D, formerly senior chemist at the Minnesota Mining and Manufacturing Company (3M). A major advantage of the process is that toxic constituents in the sewage cannot interfere with its operation. In fact, the toxic elements are incinerated and bound into a ceramic product suitable for sale as building material. The water effluent from the process meets federal secondary treatment standards without requiring biological treatment.

Let's walk through the steps of the CCBA process: First, the influent sludge is screened through 1/8-inch wedgewire for degritting. The screened materials are then mixed with a clay slurry, at a ratio of about 2 to 3 tons of clay per million gallons of water, and a gravity separator removes grit and sand from the mixture. Alum is added at anywhere from 4 to 100 parts per million (ppm), along with 1 ppm of polymer polyacrylic acid, used as a glue base. A specially designed "sheet flow mixer" is used to optimize the mixing rate without causing dissociation of the flocculent. The sheet mixer is far more energy efficient than conventional, high horsepower mixing "churners."

Finally, the floc is settled in a clarifier to form a thick sludge. The total process successfully removes 91 percent of the suspended solids and 88 percent of the biological oxygen demand (BOD) from the wastewater. The clean water decanted off the top of the clarifier can be further treated before release into the environment. Alum can achieve a 95 percent phosphate removal, and a rice hull char filter can remove 43 percent of the nitrogen.

The sludge mixture at the bottom of the clarifier is pumped into a filter and pressed into a cake, 37.3 percent of which is solid. This sludge is then mixed with more clay in a mogul mixer until a putty-like composition is achieved. The sludge-clay solid is extruded through die and cut into pellet particles. These pellets are tumbled to round the shape, and then heated in a kiln at 2,000 degrees Fahrenheit to form what is known as "low weight aggregate" (LWA). The heating process incinerates any hazardous inorganic and organic chemicals in the sludge, and traps the remaining ash in a ceramic matrix. Water and the strongest acid conditions found in nature will not leach any compounds from the ceramic pellet. These pellets, used as an aggregate in concrete production, are comparable to any on the market, and can get anywhere from $35 to $65 per ton in the Southern California area, although markets exist for LWA all over the world.

The CCBA treatment plant offers several advantages over the more conventional, biological variety. Not only does it produce a salable product out of waste, it is also simpler to design and almost a third less costly to build. A 10 million gallon-per-day CCBA facility will cost, on the average, $8 million (.80¢ per gallon of treatment capacity) as opposed to $35 million ($3.50 per gallon of capacity) for a similar-sized biological treatment plant.

Composting. The three principal methods most commonly used to manage treated sewage sludge are land application, composting, and incineration. The Catawba River Pollution Control Facility in Morganton, North Carolina, conducted a detailed survey of all three options, and concluded that for land application, composting, and incineration, the handling costs for each dry ton of sludge would be $49, $59, and $141, respectively. Although the land application option seemed to have a slight cost advantage, the officials at the Morganton facility soon learned that it was impractical because of community resistance to land application, stemming from fears of groundwater contamination and the stigma attached to sewage. Morganton officials therefore decided to implement the composting option because it offered a more dependable solution and a more marketable and manageable final product. They were the first municipality in North Carolina to develop sewage sludge composting, and as such, they have done a lot of experimentation and built up an impressive data base which other municipalities can now use.

There are currently two methods of composting available on the market: One is the older, more manual "static pile" system such as is used by the Morganton facility, and the other is an enclosed, fully mechanical composting system, which we will discuss later.

The first step of the static pile process, as Morganton uses it, consists of mixing a centrifuged sludge cake (14 to 15 percent solids) with a bark bulking agent at a ratio of 9 yards of bark to 4 yards of sludge. The bark provides free air space and ensures a continuous supply of oxygen throughout the pile. Next, perforated steel pipes are laid on a well-drained, blacktopped area and the pipes covered with a layer of bark. The sludge mixture is piled on the bark layer, to a height of 7 to 8 feet, and covered with a blanket of finished compost for insulation and odor control. Each pile represents the accumulation of 2 to 3 days of sewage. At the end of the composting process, air is drawn through the perforated pipes by blowers and then exhausted through the pile of finished compost for deodorization and sterilization. An operator measures the temperature and oxygen level in the compost pile using probes, and adjusts the blowers to maintain a 5 percent oxygen content and a minimum temperature of 140 degrees Fahrenheit for 4 days. After this drying process is completed, the pile is taken down and cured for an additional 28 days before being screened through a 1/2- or 1-inch screen to remove coarse bark. The system produces about 4.5 tons of finished compost each day.

The chemical composition of Morganton's compost is 1.67 percent nitrogen, .38 percent phosphorous, and .018 percent potassium. The material has a number of uses:

- as a potting medium or mulch;
- as a soil supplement for lawns;
- as a substitute for topsoil in land reclamation and public works projects;

- for turfgrass production;
- for nursery production of trees and ornamental plants;
- as a filler on golf courses and cemeteries;
- for the revegetation of disturbed lands (e.g., from surface mining); and
- for landscaping parks and public buildings.

In a properly controlled composting process, practically all of the common pathogenic organisms associated with sewage are destroyed or reduced to insignificant levels. Morganton's compost has been determined to be environmentally safe, and it does not emit any unpleasant odors.

Morganton invested $104,000 in their static pile system, a figure far lower than the $442,000 that would have been required for land application. They are currently selling their compost product, called "Morganite," to the general public for $4.50 per cubic yard, equivalent to about $20 per ton. Production cost for the compost is $53.75 per ton, based on annual operating and maintenance expenses. The price is purposely kept low in order to acquaint the public with the product and to eliminate any stigma attached to sewage compost. Although Morganton is not making money on the operation, their sludge management costs are far below those for other systems, including incineration and land application. Marketing of the compost product has been so successful that the city is unable to keep up with customer demand.

A second method of composting is a fully enclosed, mechanical system, such as is planned for the Hickory/Newton/Conover complex in Catawba County, North Carolina. The new complex was designed by the consulting firm of Burmeister, Wright & Associates, and construction is planned for the fall of 1986. When completed, the Hickory/Newton/Conover complex will be the first mechanical, enclosed composting system in North Carolina.

The new system will include reactors in which the actual composting process will take place; a carbonaceous storage tank; air blowers and scrubbers; a conveyor and solids handling system; and other associated buildings and structures. Sludges from several treatment plants and septage systems in Catawba county will be transported in tankers to a regional facility for dewatering before being composted.

Capital costs for the entire composting system are estimated to be about $6.1 million. Although this amount is far more than the capital investment required for a static pile system, the manual labor and health risks associated with a mechanical, enclosed system are minimal. Approximately 75 percent of the $6.1 million investment will be offset by federal grants, and the North Carolina Department of Natural Resources and Community Development will reimburse the facility for funds used to prepare project plans and specifications. The finished compost will be sold at $20 per ton, and with the facility processing at least 20 tons of dry sludge per day, the revenue generated should be approximately $400 per day, or $146,000 per year. The facility operators expect to save $827,000 per year in avoided disposal costs.

Contacts For Further Information

1. David Roche, Biologist
   Fossil Production Department
   Duke Power Company
   P.O. Box 33189
   Charlotte, North Carolina 28242
   Contributing expertise: Recovery and marketing of fly and bottom ash. Incineration of PCBs.

2. Grover Dobbins
   Carolina Power and Light Company
   Shearon Harris Energy & Environmental Center
   Route 1, Box 327
   New Hill, North Carolina 27562
   Contributing expertise: Recovery of fly and bottom ash.

3. Bill Fletcher, President
   Ash Management Corporation
   777 Franklin Road, Suite 100
   Marietta, Georgia 30067
   Contributing expertise: Marketing of fly and bottom ash (contracted by Carolina Power and Light).

4. Gerry Gordon, President
   Monier Resources
   45 Northeast Loop 410, Suite 100
   San Antonio, Texas 78216
   Contributing expertise: Marketing of fly and bottom ash (contracted by Carolina Power and Light).

5. PA Industries
   P.O. Box 2370
   Chattanooga, Tennessee 37409
   Contributing expertise: Marketing of cenospheres.

6. Russell Propst, Health Physicist
   Nuclear Production Department
   Duke Power Company
   P.O. Box 33189
   Charlotte, North Carolina 28242
   Contributing expertise: Segregation of contaminated waste streams, installation of ion-exchange system for decontamination, and upgrading of current decontamination equipment.

7. Bill Roland, President
   Natural Power
   3000 Gresham Lake Road
   Raleigh, North Carolina 27609
   Contributing expertise: System for collecting methane gas and converting it to electric power.

8. Bob Smith, Executive Director
   CCBA Resource Center (USA)
   West First National Bank Building
   St. Paul, Minnesota 55101-1315
   (612) 228-1490
   Contributing expertise: Development of the chemical coordinate bond and adsorption system for treating sewage waste.

9. Carl D. Hennessee
   City of Morganton
   P.O. Box 430
   Morganton, North Carolina 28655
   Contributing expertise: Development of the "static pile" composting system for treating sewage waste.

10. Recil Wright
    Burmeister, Wright & Associates, Consultants
    2 Parkway Plaza, Suite 170
    Charlotte, North Carolina 28210
    Contributing expertise: Implementation of a fully-enclosed, mechanical composting system for treating sewage waste.

8

Financing a Waste Reduction Program

The great majority of businesses generating hazardous wastes today can implement waste reduction strategies at a reasonably low cost and expect a quick return on their investment. Among the numerous surveys that substantiate this, studies conducted by the National Academy of Sciences characterize most of American industry as in the early, low-cost phase of waste reduction.

In one instance, a management initiative requiring no more investment than a couple of days staff time paid $129,600 yearly in real earned income and avoided costs for a zinc-smelting company. This opportunity presented itself when the smelter's management ceased to view the zinc-laden ash as a hazardous waste and instead sold it to a smelter as zinc ore.

While not every production line will find a lucrative return for such minimal investment, a management commitment will usually turn up multiple low-cost opportunities for waste reduction.

THE THREE PHASES OF WASTE REDUCTION

The implementation of a waste reduction program is usually characterized by three phases: "management initiative," "technology intensive," and "research and demonstration." Management initiative is the first phase in a company's program. It involves a perceptual shift on the part of plant personnel from the usual "end-of-pipe" disposal approach to a more integrated, system-wide emphasis on waste reduction. In this phase, waste reduction modifications stem largely from the application of strategies rather than technology. These strategies include:

- The waste audit to identify opportunities for waste reduction.
- Improved housekeeping.
- Increased operator awareness and efficiency.
- Waste stream segregation.
- An effort to involve *all* employees in seeking waste reduction opportunities.

The second and third phases are logical extensions of the first phase. As awareness of waste is heightened, new opportunities for waste reduction are recognized; however, these opportunities may now involve longer payback periods than previously.

In the technology intensive phase, processes are modified using technologies that have been identified as improving efficiency or otherwise reducing waste. These modifications can range from relatively inexpensive changes to a thorough retooling. As waste becomes an increasing expense, its reduction will become more of a focus in the engineering of new equipment and the design of new processes and products.

In the third phase, research and demonstration, a company may realize that it just doesn't have all the answers yet. Process engineers will notice that some waste streams don't respond to current strategies and technologies. A close look at these troublesome waste streams may reveal new means for their reduction.

As in any endeavor, identifying and solving such problems advances the science and results in new technologies, strategies, and products. Already, overseas competitors are marketing their solutions in the United States. Dunn and Bradstreet estimated as early as 1974 that the waste reclamation industry could become the second largest industry in the U.S. after agriculture by the year 2000. Robert Schlipp, manager of separation systems for E.I. DuPont De Nemours & Company, predicts that the sales of membrane separation products, technologies, and related systems will total $2 billion by 1995. Membrane separation of chemical compounds is only one of many technologies currently being employed to reduce waste at the source.

THE PUSH TOWARD WASTE REDUCTION

As is the case with many desirable ventures, waste reduction projects require justifications for their expense—regardless of the pay-back period. After all, people argue, if a waste reduction project entails significant financial planning, and if we already have an adequate disposal arrangement, why should we go to all the trouble? There are at least five strong justifications for taking this trouble:

1. A heavy liability follows hazardous waste disposal, and the risk of exposure is becoming increasingly difficult and expensive to insure. Superfund cleanup liabilities are already reaching deeply into corporate pockets.
2. Disposal options are decreasing due to tighter regulations, such as RCRA's phased-in ban on land disposal of hazardous waste.
3. Methods of disposal, such as landfilling, are becoming increasingly expensive for the generator, approaching as much as $300 a drum in some cases.
4. Industries are finding that more efficient resource use, recycling, and conservation can greatly reduce their raw material costs, cut down on disposal costs, and actually enhance profitability.
5. Decreased employee exposure to toxic and hazardous materials.

All five of these considerations indicate that the sooner hazardous waste generators implement waste reduction strategies, the quicker savings will accrue from avoided disposal costs and CERCLA liabilities. Moreover, aside from avoiding costs, the reduction of hazardous waste can generate real profits when valuable raw materials are used more effectively and never enter the waste stream, or are recovered from the waste stream for reuse.

DIFFERENT GROWTH STAGES MEAN DIFFERENT STRATEGIES

Businesses, like living organisms, have life cycles. The Abernathy-Utterback model of industrial life cycles proposes that industries have three stages of growth:

- the initial fluid stage, characterized by high growth and innovation
- a transitional stage, in which the company is moving toward more specialized products and services
- the mature stage, in which the company has settled into its particular "niche"

The stage that a company is in can determine the ease with which it is able to implement waste reduction. Several behavioral traits of these stages will figure in the company's decision:

- size
- the extent of capital investment and its age/amortization status
- financial strength
- growth rate, and
- management structure

Size. This is arguably the most important aspect to be considered in implementing waste reduction strategies. Key resources, such as research facilities and environmental and engineering expertise, often depend on how big the company is; logically, only very large or multiple production facilities can support such specialized departments. Large industries have other special advantages. For example, a large, diversified operation can sometimes alter process "z" so that it will use a by-product from another process, thus utilizing the waste stream from that process. A large firm can also absorb the cost of a major retooling for waste reduction if it decides to sell corporate bonds or stock to finance it.

In contrast, a small "mom and pop" shop may not even have the personnel or time to learn about waste reduction opportunities. Similarly, they often do not have the capital holdings needed to mortgage or assure loans. Usually the company is privately held, which would mean that it would be unable to issue stock as a finance mechanism.

Equipment. A company's current investment in process equipment and the age and specificity of such equipment will influence its decision on further investment. The extent of these new investments will depend largely on what stage of growth the industry is in. Businesses in transition between the early, "fluid" and later, "mature" stages have the best opportunity to capitalize on new waste reduction modifications. This is because they have defined their niche within the industry, and thus are moving from a generalized production line to more specialized equipment for their defined product. Companies in the mature stage seeking to replace worn and outmoded equipment have the same opportunity.

Most importantly, the capital investment in new equipment will allow for waste reduction technology to be designed into the product and the processing machinery. Naturally, it is far more efficient and economical to integrate waste management at the design stage than to go back and retrofit. Therefore, waste reduction should play a major role in the selection of new equipment.

Financial Strength. The financial strength of a company will affect its ability to secure additional funds for waste reduction modifications. Important measures of financial strength are:

- short-term debt (ratio of current assets to liabilities)
- the ratio of debts to total assets, and
- profitability (profit margin on products, return on net worth, and total assets).

Potential lenders and investors will ascertain a business "risk" based on its capital structure and its debt/profitability. The stage of development that a business is in will determine what form of financing needs to be sought.

In the early stages of a business' growth, funding will usually be in the form of venture capital and government assistance, such as research and development grants and small business loans. As industries mature, capital will come in the form of commercial loans, corporate bonds, and stocks. Mature industries have less need of external funds, as they are better equipped to generate cash internally.

Growth Rate. A company's growth rate affects the availability of financing. Rapid growth—the real rate of increase in a company's earnings and assets—attracts equity financing and helps justify new issues of stock. This condition, however, usually means that fiscal reserves have to be tapped to cover the expenses associated with increasing sales. This condition, then, presents both an opportunity and a restriction for waste reduction modifications.

In contrast, cash reserves are often available to a profitable company with a low growth rate. However, if a company's cash reserves are insufficient to retool (if such a major modification is feasible), then the low-growth company may have a harder time obtaining external funding than its high-growth counterpart.

The early, "fluid" stage of a company's life cycle is often characterized by high growth. Typically, product innovation and diversity are the rule. Manufacturing equipment is often generalized, and thus flexible but not optimally efficient. Although the opportunities for introducing waste modifications exist, the importance of maintaining flexibility in product design may argue against it.

On the other hand, the more mature company will know what products it should be emphasizing, and will have begun to look at enhancing production efficiency. It will start replacing marginal products and outmoded equipment with more specialized components; waste reduction should play a central role in the company's process modification decisions.

Management structure. This characteristic can dominate all others, becoming the key contribution to the success or failure of a comprehensive waste reduction program. In companies with a strong centralization of power at the CEO level, it is critical that the waste reduction program be promoted and supported from the top. In companies with a greater diffusion of responsibility, as is the current trend, management recognizes that waste reduction must be a team effort. This allows employees at all levels to contribute to a comprehensive waste reduction system.

YOU *CAN* GET THERE FROM HERE

Before examining some specific strategies for raising funds to finance the more extensive waste reduction modifications, it is necessary to first examine the conditions requiring this investment.

In the decade after World War II, the cost of many resources declined, and manufacturers were able to externalize waste disposal costs into the environment. But the falling prices of resources and the cheap price of waste management and energy that fueled the strongest economy in the world's history no longer exist. Market and regulatory forces are making resources and waste management more expensive. Energy prices are erratic, but will likely increase in the long run, despite the drop in early 1986.

Companies that successfully make the transition from past conditions to current ones will effectively apply new information and efficiently utilize their resources, including capital, energy, human resources, and raw materials. Waste is a *luxury* that competitive businesses can no longer afford. However, capital for waste reduction modifications may be a *necessity* some companies cannot afford either.

In such cases, new sources of investment capital may come from industry pools, state government programs in waste reduction, and even the communities where plants are located. Don't overlook the interests that host communities hold in a clean environment and job security. Numerous communities have taken positive action to raise money for industrial projects, because these projects were perceived as being in the public interest. Local banks may be willing to make loans to local industries to keep them competitive and to promote a healthier environment. If this argument fails to secure a loan, however, the applicant might consider enlisting the help of the local environmental group in order to strengthen its case to the bank. Although in the past, alliances between industries and environmental groups might have been unthinkable, today, waste reduction can be as attractive to business managers as it is to environmental activists.

FINANCIAL ASSISTANCE FROM GOVERNMENT

This section addresses government's role in industry's transition to waste reduction. David Anzures of California Technical Plating put forward the idea that government has "a moral obligation to help those that are being regulated." Many believe that government could best promote waste reduction if it "got out of the way"—removed the barriers and erased the disincentives. Government now exacts an expensive regulatory system on industry in an effort to *control the pollution* of the natural environment. Any modifications to this system that would reduce the cost to the taxpayer and simultaneously improve the ability of industry to meet its objectives is desirable. The reduction of waste at the source is an obvious improvement over the costly and burdensome emphasis on pollution control adopted by most state and federal regulatory agencies.

Unfortunately, governments, in particular the federal government, have been slow to respond.

Federal Government

There is no financial assistance available for the adoption of waste reduction on the federal level. True, Section 103 of the IRS tax code provides tax incentives for pollution control bonds. But there's a catch —a clause that requires the treatment or storage of an existing pollutant. This "realized waste" clause means that equipment or modifications that actually reduce or eliminate the production of pollutants are not eligible for tax-exempt financing. The Resource Conservation and Recovery Act (RCRA) prohibits the issuance of tax-free industrial bonds to fund any equipment or modifications which would bring a business into compliance with RCRA. There are similar disincentives which discourage recycling and the use of waste exchanges. For example, under RCRA, certain on-site waste recycling practices require a "part B" authorization from the EPA before the site can become a treater, storer, or *recycler* of hazardous waste. According to many industry leaders, a part B authorization is very costly and discourages many companies from engaging in waste reduction. In light of the 1984 amendments to RCRA, which require that hazardous waste generators prepare and implement plans for waste reduction, these regulatory disincentives are obviously counterproductive to RCRA's stated objective.

State Government

Prompted by citizens and local businesses, state governments have been much more responsive to industries' financial needs with regard to waste reduction. State governments are providing a variety of waste reduction incentives, albeit on an inadequate scale.

In states where no such program exists, a pioneering effort by enlightened businessmen to introduce financial assistance for waste reduction in their states will greatly contribute to the development of this management strategy for industrial by-products. Thus, state lobbyists may want to look at these four items as an initial shopping list:

- Technical Assistance
- Grants
- Loans
- Tax Incentives

Technical assistance is an indirect form of financial assistance, while grants, loans, and tax incentives are actual funding strategies for implementing waste reduction modifications. As strategies, they are nothing new; however, their use for waste reduction purposes are new, and in many cases, unfortunately, underutilized.

Technical Assistance. Technical assistance and information dissemination are indirect forms of financial assistance. Many states will bear the cost of developing a data base and of training personnel to assist industries in developing waste reduction programs. This function is especially valuable for smaller businesses which lack the resources to research and develop their own programs.

Technical assistance is disseminated through a variety of environmental, economic, or administrative agencies. Some programs operate independently, or as an adjunct to a university program or state waste management advisory board. Typically, they are engaged in three primary tasks:

- the identification and collection of technical information useful to the local industry
- preparation of appropriate informational material
- the dissemination of this information

These three functions are performed in various ways. Many programs publish a newsletter listing new publications pertinent to waste reduction, or they might list waste streams for exchange (see related section in chapter 2, "What Is Waste Reduction?").

Conferences and workshops are also used for sharing information. Conferences usually focus on the basic principles of waste reduction and also provide specific industrial examples that illustrate the basic principles. In the course of the conference, industrial leaders will be informed of the various services available through the state's technical assistance program. Workshops, on the other hand, are designed to transfer industry-specific technology to a specialized audience. Some programs have sufficient technical staff to provide on-site evaluation and consultation to individual companies.

A few states have committed resources to the advancement of our current knowledge about waste reduction technology and strategy. Research projects conducted by state programs include:

- company surveys of the major factors influencing waste management decisions
- case studies of companies that have realized the economic and environmental benefits of waste reduction technologies
- technical manuals on waste reduction opportunities in specific industries
- funding for specialized research to develop new waste reduction technology for extremely hazardous, high-volume waste streams

These programs, often called "Pollution Prevention Pays," have been implemented by a number of states, including North Carolina, New York, Minnesota, Illinois, Georgia and Massachusetts. In addition, Environment Canada funded a demonstration technical assistance project in the Region of Waterloo, Ontario. The project was designed as a prototype to be duplicated in other parts of Canada, and as such, there is good documentation of this program available (see References and Resources).

Grants. Grant programs to finance waste reduction are particularly scarce, as is the case whenever money is being given away. Currently, such programs exist only in Canada, Wisconsin, Michigan, and North Carolina. Generally, the recipient of the grant must match the funds it receives, either with money or in kind. Funding can vary from $5,000 to as much as $30,000 per industrial applicant.

The goal of a grant program is to promote the development and implementation of waste reduction technology through a number of approaches:

- information dissemination
- environmental/waste audits
- technology demonstration
- design and implementation of equipment or facilities

The actual reduction of waste is often an outcome; however, in the information dissemination and waste audit areas, the recipients are in an early stage of planning their waste reduction modifications. In such cases, the groundwork for reduction is laid, while the actual reduction may not be realized until a later time. Similarly, in the study of new waste reduction technologies, there are no guarantees of success. The money is awarded simply to research the *feasibility* of adopting a new waste reduction method; there are no allowances for actually installing equipment.

In contrast, other grants are made to individual businesses that will install a specific technology as a demonstration project for the rest of the industry. Other industries can fully expect to reap the rewards of waste reduction. The grant money is given usually with the stipulation that the results of the project be made available to all interested parties.

Loans. Like grant programs, state-aided loan programs for waste reduction are rare and need state champions to secure funds. New York and Massachusetts are considering providing low-interest loans to certain industries, and the idea has been advocated in Maryland.

Most states offering or considering offering loan programs require that companies meet three qualifications before they can become eligible to receive a loan:

- The applicant must be a small-to-medium sized business.
- The applicant must be unable to secure a conventional bank loan and must not have the capital to finance the waste reduction project on its own. (Many supporters of state-offered loans believe that they should be targeted specifically to businesses that are unable to secure private loans.)
- The applicant must be categorized in a standard industrial classification that has a high potential for waste reduction.

Funding for such loan programs comes from a number of sources, including legislated bond issuances, revenues from state general funds, and revenues from a hazardous waste-end tax. Another source could be tax-exempt industrial bonds, such as are used by facilities that treat "realized pollution." However, the Internal Revenue Service says that waste reduction technology does not treat realized pollution and therefore is not eligible for financing through this method. This policy reflects the older emphasis on pollution control and must be revised to reflect a new emphasis on waste reduction. The business community may want to address this area with state and federal authorities if they wish to increase the incentives for waste reduction.

Other loan-associated strategies include loan guarantees, and a state subsidy to mitigate or eliminate interest on a loan (this would actually qualify as a grant). A loan guarantee program, in which the state would assume responsibility for the loan in case the company defaults, is an inexpensive means for states to assist industry in financing the more capital-intensive waste reduction programs. Assuming that the agency overseeing the program is careful in its assessment of potential projects, this means of financial assistance could be the most effective in providing capital to companies for waste reduction.

In contrast, interest subsidies will most likely be insufficient to stimulate industrialists to initiate reduction projects; they would only serve to "reward" companies investing in such modifications.

Tax incentives. Tax incentives can take two forms:

- Income subject to tax is reduced by an amount equal to the cost of the waste reduction project.
- The taxes themselves are reduced by a percentage of the equipment cost.

As noted earlier, pollution *control* equipment can be written off federal taxes, although waste reduction or waste elimination equipment does not currently qualify. This same condition exists in nearly every state which grants a similar incentive. Wisconsin modified its law to include waste reduction technologies in the exemption.

Some states require that waste reduction equipment be certified as such by a designated agency before it can be exempted from taxation.

One such state is North Carolina, which offers numerous opportunities for waste reduction deductions. For example, the cost of facilities and equipment may be deducted from the company's state income tax; from individual taxes; from the corporate franchise tax; and from capital stock, surplus, and undivided profits.

All four forms of financial assistance—technical assistance, grants, loans, and tax incentives—can play an instrumental role in an industry's decision to invest in waste reduction technology. Grants, loans, and technical assistance can overcome the "tight money" obstacle. Tax incentives and loan interest subsidies, on the other hand, are thought by some to be not substantial enough to help an interested company overcome its financial obstacles. Rather, they provide an incentive or reward for desired behavior. Given that most industries are in the early, low-cost phase of waste reduction, perhaps the first step of any state program should be to make companies more aware of waste reduction strategies and technologies and to help them implement such opportunities.

PRIVATE SECTOR INITIATIVES

In the absence of any government assistance as described above, several innovative groups have pioneered efforts in the private sector which accomplish the same objectives. The following two examples illustrate cooperative ventures designed to serve a regional industry, in this case the metalplating industry. One venture was championed by the industry's trade association, while the other was coordinated by a metropolitan council.

The opportunity for the projects arose out of the new pretreatment standards imposed by the Clean Water Act of April, 1984. These standards require the removal of heavy metals from the waste stream prior to its discharge into the local sewer. Many small electroplaters have already been put out of business by this regulation, and many more are threatened. The EPA estimates that the Twin Cities of St. Paul/Minneapolis in Minnesota could lose up to 20 percent of its electroplating industry if these companies are forced to purchase conventional sludging technology in order to meet regulations. Consequently, nearly all electroplaters are facing a simultaneous need for improved waste management.

In 1979 and 1981, the EPA investigated the feasibility of a centralized processing and recycling facility using ion-exchange technology, or similar filtration technology, to serve the needs of participating plants. Each plant would have ion-exchange equipment installed to remove metals from the effluent prior to its release into the sewer system. Then, the plant's ion-exchange resins would be "regenerated" at the central facility, and the metals collected and recycled.

Aside from providing a substantially lower cost in treating waste over current methods of landfilling, a central recycling facility, together with the industries' on-site ion-exchange treatment, presents numerous other advantages:

- It is the most effective technology to prevent the discharge of toxic metals into sewer systems (or surface waters). This will reduce the heavy metal concentrations in sewage sludge and render it more suitable for composting and agricultural land spreading.
- It will allow the reuse of industrial process water, which leads to lower water and sewer bills.
- It will attract and retain high technology corporations by allowing them to remain healthy and competitive.

In New Jersey, the Master's Association of Metal Finishers (MAMF) established a non-profit arm, the Research Foundation, to assess the feasibility of a centralized facility for treating electroplating waste in the metropolitan New York area.

The Foundation is currently finalizing arrangements for siting the facility and for purchasing the equipment, and has secured funds from a number of sources to conduct the necessary research and planning required for the project. Ninety members of the Master's Association have contributed $2,500 each, and a variety of grants and other assistance where obtained from the EPA, the New Jersey Department of Environmental Protection, the New York City Office of Economic Development, the Port Authority of New York and New Jersey, the

Publicly Owned Treatment Works in New Jersey and New York, and the New York Community Trust.

Conventional chromium reduction, cyanide reduction, and precipitation systems would cost each electroplating shop an average of $200,000 to purchase and install. This is far beyond the means of many small job shops which dominate the industry in the New York-New Jersey metropolitan region.

In contrast, the proposed project sponsored by the Research Foundation will save each company $14,562 per year in operating costs over their current system. As one consultant put it, "This presents an unprecedented opportunity for electroplating companies to come into compliance without facing a price increase to pay for compliance. A distinct competitive advantage!"

Each participating company will install an ion-exchange recovery system in its shop. When the ion-exchange resins are full, they will be brought to the Central Recovery Facility (CRF), where the metals will be removed and sold to industries which can use them in their manufacturing processes. The CRF's fee for handling the metal waste is much lower than that for conventional disposal methods, because each plater is credited with a dollar amount equal to the value of his reusable metal. The consultants report that in one year, nearly 50 percent of the participating companies will pay off their initial investment in ion-exchange technology. The total cost for the CRF is estimated to be approximately $3.6 million; however, the facility should pay for itself in 37 months. In addition, the facility should generate a net revenue of $685,000 per year from the sale of the recovered metal.

The Twin Cities area in Minnesota, like metropolitan New York, is home to a large sector of metal using and producing industries. The same wastewater pretreatment standards which prompted the Master's Association to promote a CRF in New Jersey also stimulated a similar effort in Minnesota.

The Metropolitan Council, the Twin Cities' regional planning organization, appointed a task force composed of members from the regulatory agencies and the industry trade associations in order to coordinate an industry-wide compliance with the new wastewater requirements. Members of the industry wanted to comply in the most economical, but still environmentally sound, way possible.

The task force's job was threefold:

- to determine potential markets for a centralized treatment and metal recovery facility
- to identify any companies likely to participate
- to characterize the waste products coming into the facility

The study was funded by industry donations and a grant from the Minnesota Economic Development Section. Upon completion of the study, industry representatives and trade association officers investigated the feasibility of the task force's recommendations. An industry-dominated corporation was set-up to finance and develop the proposed central recovery facility.

In 1983, a site was purchased, and in 1984, ion-exchange technology was tested with the cooperation of the University of Minnesota. In 1985, the required permits and licenses were applied for.

The capital investment required for the CRF is approximately $7.5 million: $2.5 million is allocated for building and improvement, $3.5 million for equipment, and $1.5 million for resins and canisters. A pooled industrial development bond offering, sponsored by the Minneapolis Community Development Agency, will be made to customers in order to fund the investment.

Like the New Jersey project, the Minnesota plan will allow industry to comply with regulations without having to substantially increase their current operating costs.

CONCLUSION

Traditionally, our society has approached pollution by adding end-of-pipe pollution controls and by disposing with an "out-of-site, out-of-mind" perspective. Waste reduction requires a new way of looking at things, with new technological angles to realize the objectives. Anyone who wants to benefit from the current techniques must first be able to understand them.

The financing of waste reduction is much less an obstacle than overcoming the initial reluctance to alter the old way of doing things.

States could overcome the inertia by offering financial assistance and incentives for industry to look at waste reduction.

In the absence of government incentives or assistance, many industries, singularly or cooperatively, have found innovative strategies for securing funds from organizations and agencies that endorse waste reduction as a strategy for waste management. A promising source of funds is the community in which a waste generator is located. The collective assets of a community are considerable. As an alternative to private banks, communities can make low-interest loans to companies interested in waste reduction, and companies can usually repay these loans with the savings and avoided costs generated by the waste reduction project. This community-company arrangement can help improve the local environment, improve the occupational health of workers, and keep the local industry competitive—thus preserving jobs. Management commitment and technical training in waste reduction techniques for the engineering staff are the most critical resources now required to make waste reduction a reality.

The products of the future will require greater information input and less energy, materials, and labor. Companies that reflect this trend in their production will be better prepared to remain competitive. Given the increasing production costs associated with managing waste, investments in the knowledge needed to reduce these costs is an important factor in keeping a company profitable.

References and Resources

Adamson, V. *Breaking the Barriers: A Study of Legislative and Economic Barriers to Industrial Waste Recycling*. Ontario, Canada: Canadian Environmental Law Research Foundation and the Pollution Probe Foundation, 1984.

Campbell, M. and W. Glenn. *Profit from Pollution Prevention*. Toronto, Ontario: Pollution Probe Foundation, 1982.

Environment Canada. *Waste Abatement, Reuse, Recycle, and Reduction Opportunities in Industry*. Kingston, Ontario: Environment Canada, 1984.

Government Institutes. Proceedings of seminar, "Hazardous and Solid Waste Minimization." Washington, D.C.: Government Institutes, 1986.

Huisingh, D., and V. Bailey. *Making Pollution Prevention Pay*. New York, N.Y.: Pergamon Press, 1982.

Kohl, J., and B. Triplett. "Managing and Minimizing Hazardous Waste Metal Sludges." Raleigh, N.C.: North Carolina State University Industrial Extension Service, 1984.

Kohl, J., P. Moses, and B. Triplett. "Managing and Recycling Solvents." Raleigh, N.C.: North Carolina State University Industrial Extension Service, 1984.

McRae, George. "Enviroscope: In-Process Waste Reduction" (3 parts). *Journal of the American Electroplaters' Society*, 1985, Vol. 72, Nos. 6, 7, and 8.

Noll, K., C. Haas, C. Schmidt, and P. Kodukula. "Recovery, Recycle, and Reuse of Industrial Wastes." Chelsea, MI: Lewis Publishers, 1985.

Overcash, M. *Techniques for Industrial Pollution Prevention.* Chelsea, MI: Lewis Publishers, 1986.

Sarokin, D., W. Muir, C. Miller, and S. Sperber. *Cutting Chemical Wastes.* New York, N.Y.: Inform, Inc., 1985.

Tavlarides, L. "Process Modifications for Industrial Pollution Source Reduction." Chelsea, MI: Lewis Publishers, 1985.

Technical Insights, Inc. *New Methods for Degrading/Detoxifying Chemical Wastes.* Englewood/Fort Lee, N.J.: Technical Insights, Inc., 1986.

U.S. Congressional Budget Office. "Hazardous Waste Management: Recent Changes and Policy Alternatives." Washington, D.C.: Superintendent of Documents, U.S. Government Printing Office, 1985.

For More Information . . .

ALSO AVAILABLE FROM THE INSTITUTE FOR LOCAL SELF-RELIANCE

Resource Recovery State-of-the-Art: A Data Pool for Local Decision-makers, (November 1985, $36.50; an up-to-date overview of waste recovery and utilization technologies with comparisons of systems, economics and environmental impacts).

New City-States, (1982, $7.00; describes how America's cities are looking inward for solutions to their problems).

Waste-to-Wealth: A Business Guide for Community Recycling Enterprises, (1985, $36.50; a how-to guide for preparing business and investment plans for recycling enterprises. Case studies include descriptions of industry, equipment and equipment suppliers, market survey techniques, sample cash flows, and capital requirements).

Upcoming reports include a second volume of hazardous waste minimization case studies, an overview of the comparative costs of various garbage disposal technologies, a book on alternatives to large-scale incineration plants, and a report on dioxin/furan leachate from incineration ash.

Further information and a publications list are available from the Institute for Local Self-Reliance, Department A, 2425 18th Street, N.W., Washington, D.C. 20009, (202) 232-4108.